Bio-diesel

About the Author

Dr. G. Ravi Kiran Sastry is a Professor in the department of Mechanical Engineering in Chaitanya Engineering College, Visakhapatnam of Andhra Pradesh. Dr. .Sastry did his M. Tech (Thermal Engineering) in Jawaharlal Nehru Technological University, Hyderabad and was awarded his PhD from Andhra University, Visakhapatnam with a specialization in alternative fuels for I.C. Engines.

Dr. Sastry has huge teaching as well as Research experience. Dr. Sastry is an ardent researcher in the area of finding alternative fuels as a replacement to commercial fuels especially biodiesels of non-edible oils such as Jatropha, Mahua and Palm Kernel oils.

Dr. Sastry has a good number of research paper contributions in national as well as international journals/conferences in the thrust area of the performance of D.I. Diesel engine using Biodiesels.

Bio-diesel
biodegradable alternative fuel for diesel engines

Dr. Gadepalli Ravi Kiran Sastry

Readworthy
New Delhi

First published 2008

Readworthy Publications (P) Ltd.

Regd. Off.: A-18, Mohan Garden
Near Nawada Metro Station
New Delhi – 110 059 (India)
Phone: 011-2537 1324

Branch: 4662/21, Ansari Road
Daryaganj
New Delhi – 110 002
Phone: 011-43549197

Fax: +91-11-2537 1323

Email: info@readworthypub.com
Web: www.readworthypub.com

Cataloging in Publication Data–DK
Courtesy: D.K. Agencies (P) Ltd. <docinfo@dkagencies.com>

Sastry, Gadepalli Ravi Kiran, 1972-
Investigations on bio diesels : a permanent replacement to petro diesels / Gadepalli Ravi Kiran Sastry.

p. cm.

Includes bibliographical references (p.)
Includes index.

ISBN 13: 978-81-89973-50-6 ISBN 10: 81-89973-50-9

1. Biodiesel fuels--Research. 2. Diesel motor --Alternative fuels. 3. Diesel motor--Vibration. 4. Diesel motor--Noise. I. Title.

DDC 662.669 072 22

Printed at Salasar Imaging Systems, Delhi - 35

Foreword

Depletion of petroleum products may not be taking place within offing because earth's reserves are immense. But cost of exploration may escalate and finally the petroleum products may be dearer day-by-day. Already, the day has come the cheaper diesel of erstwhile times has become unbearably costlier. In this context, a viable and sustainable alternative fuel is necessary to cater to the needs of mind boggling fleet of automobiles in the world. It calls for the necessity to decrease the pollution levels also complying with the emission norms to keep the environment clean. The advent of biodiesel has come to the rescue in this warranting situation. This fuel is renewable originating from the vast plant population of the globe. Basically, this oil is biodegradable and controls the greenhouse gas which is responsible for the global warming. Department of Marine Engineering, AU College of Engineering, is one of the pioneering departments in the country which can boast of diversifying research in the area of biodiesel applications and alternative fuels. It has pioneered many studies in this field with the concerted effort of researchers in designing the systems to produce biodiesel in the laboratory at the economically viable rates and implement them in running the diesel engines in lieu of petro-diesel. Much accuracy is called for in the measurement of the engine parameters which can define exactly the engine behaviour with the new trial. Dr. Sastry, worked hard in streamlining the systems to design biodiesel plants to bring succour to the community of the diesel users. Dr. Sastry strove to exploit non-edible oil plants to replace diesel oil usage in the conventional diesel engines. The day is not far away to see that all diesel engines run by petro-diesel will be replaced by biodiesel, partially or fully.

Dr. B.V.Appa Rao
Department of Marine Engineering
A.U College of Engineering
Andhra University
Visakhapatnam-530 003

Acknowledgements

I express my deep sense of gratitude to **Dr. B.V. Appa Rao**, Professor, Department of Marine Engineering, A.U. College of Engineering, for his expert guidance and support during the course of this study.

I express my gratitude to Dr. T. V. K. Bhanu Prakash, Professor and Head of the Department of Marine Engineering, for his continuous encouragement.

I also extend my heartiest thanks to the entire faculty of Marine Engineering, of A. U. College of Engineering for their continuous encouragement; Prof. K.V.L. Raju, Principal, M.V.G.R. College of Engineering, for this cooperation and support; and Prof. R.N. Someswara Rao, formerly at A.U.C.E. for this valuable suggestions.

Finally, I take this opportunity to say big thank you to my mother, wife and children for their sincere cooperation.

Contents

List of Figures

List of Tables

Nomenclature

α	:	Equivalence ratio
φ	:	Relative Equivalence Ratio
γ	:	Ratio of Specific Heats
η	:	Efficiency
μ	:	Dynamic Viscosity
ν	:	Kinematic Viscosity
θ	:	Crank Angle
ρ	:	Density
λ	:	Wavelength
τ	:	Shear Stress
$\eta_{b.th}$	:	Brake Thermal Efficiency
$\eta_{I.th}$	:	Indicated Thermal Efficiency
η_m	:	Mechanical Efficiency
ΔP	:	Pressure Difference
ΔT	:	Temperature Difference
ABDC	:	After Bottom Dead Center
ASME	:	American Society of Mechanical Engineers
ASTM	:	American Society for Testing and Materials
ATDC	:	After Top Dead center
B.T.E., b.th.η	:	Brake Thermal Efficiency
B100	:	Pure Biodiesel
B2	:	Blend of 2% Biodiesel and 92% Petroleum Diesel
B20	:	Blend of 20% Biodiesel and 80% Petroleum Diesel.
BBDC	:	Before Bottom Dead center
BDC	:	Bottom Dead center
bmep	:	Brake Mean Effective Pressure
bp	:	Brake Power
bsfc	:	Brake Specific Fuel Consumption
BTDC	:	Before Top Dead center
CA	:	Crank Angle
CAD	:	Crank Angle in Degree
CHRR	:	Cumulative Heat Release Rate
CO	:	Carbon Monoxide
D.I.	:	Direct Injection

deg	:	Degree(s)
ECA	:	Engine Cycle Analysis
F.C.	:	Fuel Consumption
F.I.A.	:	Fuel Injection Analysis
FME	:	Frying Oil Methyl Ester
HC	:	Hydro Carbon
HHV	:	Higher Heating Value
HSDI	:	High-Speed Direct Injection
I.T.E., i.th.η	:	Indicated Thermal Efficiency
I.D.I.	:	Indirect Injection
imep	:	Indicated Mean Effective Pressure
ip	:	Indicated Power
isfc	:	Indicated Specific Fuel Consumption
JME	:	Jatropha Oil Methyl Ester
LHV	:	Lower Heating Value
LOME	:	Linseed Oil Methyl Ester
MME, MOME	:	Mahua Oil Methyl Ester
NHRR	:	Net Heat Release Rate
OEMs	:	Original Engine Manufacturers
PKME	:	Palm Kernel Oil Methyl Ester
POME	:	Palm Oil Methyl Ester
PORIM	:	Palm Oil Research Institute of Malaysia
RME	:	Rapeseed Oil Methyl Ester
rpm	:	Revolutions Per Minute
SAE	:	Society of Automotive Engineers
SEC	:	Specific Energy Consumption
sfc	:	Specific Fuel Consumption
SME	:	Sunflower Oil Methyl Ester
TCSBO	:	Thermally Cracked Soyabean Oil
TDC	:	Top Dead Centre

Names of the Oils Used in this Study

S.NO	**Common Name** *(Local language names are given in the parenthesis)*	**Botanical Names** *(Family Name is given in the parenthesis)*
1.	JATROPHA (***Ratanjyot,Nepalam, Adavi Amudam,Undigapu,Physic Nut,Purging Nut,etc.***)	Jatropha Curcas Linn. ***(Euphorbiaceae)***
2.	MAHUA ***(Madhuca,Mowra,Ippa, illupal, etc.)***	Madhuca Indica,Bassia LatifolaRoxburghi. ***(Sapotaceae)***
3.	PALM ***(Oil Palm, Khajuri)*** (***P.s.***:Palm Kernel is Obtained from Oil Palm)	Elaeis guineenis, Elaeis oleifera. ***(Palmae)***

1

Introduction

Biodiesel is one of the widely tested alternative fuels in the market. A number of studies have been done and the results show the performance of biodiesel similar to that of petroleum diesel. It is also beneficial to human health and environment when compared to diesel. The research includes studies performed by the U.S. Department of Energy, the U.S. Department of Agriculture, Stanadyne Automotive Corp. (the largest diesel fuel injection equipment manufacturer in the U.S.), Lovelace Respiratory Research Institute, and Southwest Research Institute. Biodiesel is the first and only alternative fuel to have fulfilled the rigorous health effects testing requirements of the Clean Air Act. Biodiesel has been proven to perform similar to diesel in more than 40 million successful road miles and in virtually all types of diesel engines, countless off-road miles and countless marine hours. Currently more than 100 major fleets use biodiesel

One of the major advantages of biodiesel is that it can be used in existing engines and fuel injection equipment with little impact on the operating performance. Biodiesel has a higher cetane number than U.S. diesel fuel. In more than 30 million miles of in-field demonstrations, B20 showed similar fuel consumption, horsepower, torque, and haulage rates as conventional diesel fuel. Biodiesel also has superior lubricity and has the highest BTU content in the comparison to any other any alternative fuel (falling in the range between #1 and #2 diesel fuel).

Biodiesel becomes gel in very cold temperatures, just as the common #2 diesel does. Although pure biodiesel has a higher cloud point than #2 diesel fuel, typical blends of 20 per cent biodiesels are managed with the same fuel management techniques as #2 diesel. Blends of 5 per cent biodiesel and less have virtually no impact on cold flow.

Biodiesel can be operated in any diesel engine with little or no modification to the engine or the fuel system. Pure biodiesel (B100) has a solvent effect, which may release deposits accumulated on tank walls and pipes from previous diesel fuel use. With high blends of biodiesel, the release of deposits may clog filters initially, therefore, precautions should be taken to replace fuel filters until the petroleum build-up is eliminated. This issue is less prevalent with B20 blends, and there is no evidence that lower-blend levels such as B2 have caused filters to plug. The use of a 2 per cent blend of biodiesel, it is estimated, increases the cost of diesel by 2 or 3 cents per gallon, including the fuel, transportation, storage and blending costs. Any increase in cost will be accompanied by an increase in diesel quality since low-blend levels of biodiesel greatly enhance the lubricity of diesel fuel.

The recent switch to low-sulphur diesel fuel has caused most Original Engine Manufacturers (OEMs) to switch to components that are also compatible with biodiesel. In general, biodiesel used in pure form can soften and degrade certain types of elastomers and natural rubber compounds over a period of time. Using high per cent blends can impact fuel system components (primarily fuel hoses and fuel pump seals) that contain elastomer compounds incompatible with biodiesel, although the effect is lessened as the biodiesel blend level is decreased. Experiments during the last seven years with B20 has show that no changes to gaskets or hoses are necessary.

The biodiesel industry has been active in setting standards for biodiesel since 1994 when the first biodiesel taskforce was formed by the American Society for Testing and Materials (ASTM). ASTM approved a provisional standard for biodiesel (ASTM PS 121) in July 1999.

Most fuel today is used much before six months, and many petroleum companies do not recommend storing petroleum diesel for more than six months. The current industry recommendation is that biodiesel be used within six months, or reanalyzed after six months to ensure the fuel meets ASTM specifications (PS 121-99). A longer shelf life is possible depending on the fuel composition and the use of storage enhancing additives. The use of biodiesel in existing diesel engines does not void parts and materials workmanship warranties of any engine manufacturer.

There are thirteen companies that have invested millions of dollars into the development of the biodiesel manufacturing plants actively marketing biodiesel. Based on existing dedicated biodiesel processing capacity and long-term production agreements, more than 200 million gallons of biodiesel capacity currently exists. Many facilities are capable of doubling their production capacity within 18 months.

The U.S. Department of Agriculture announced in January 2001 the implementation of the first programme providing cost incentives for the production of 36 million gallons of biodiesel. Bills supporting the use of biodiesel and ethanol were also introduced in the U.S. Congress in 2001, including one that would set a renewable standard for fuel in the U.S. and one that would give biodiesel a partial fuel excise tax exemption. More than a dozen states have passed favourable biodiesel legislation.

Biodiesel is a non-polluting recycled fuel made from organic oils. It is chemically called Free Fatty Acid Methyl Ester. It is made from processed organic oils and fats, and can be burned in normal diesel engines just like normal mineral diesel, but its use neither pollutes the atmosphere nor adds to the causes of global warming. It is also possible to make good bio-diesel from waste vegetable oils like used chip fat. In this way, burning bio-diesel turns a waste disposal problem into a non-polluting fuel source.

Organic fuels are derived from plant and animal fats. Mineral fuels are derived from the fossil remains of decomposed organic matter extracted from below the surface of the earth. It is a common knowledge that the resources of mineral oils are nearly depleted, and the cost of extracting the remaining reserves will become increasingly high. There is an urgent need to find other sources of energy before mineral fuel supplies run dry.

It is also well known that burning of fossil fuels increases the level of carbon-dioxide in the atmosphere as the carbon locked within the earth's crust, when burned, is released into the atmosphere as exhaust gasses. This is the main cause of the 'Green House' effect in which the overall temperature of the globe increases as it becomes enveloped within a pool of carbon dioxide. This process is believed to be the main cause of global warming, which is now a well-accepted fact even amongst those who were most skeptical. All the time we burn normal petrol or mineral diesel, we are therefore actively contributing to global warming.

However, the burning of organically derived fuels does not contribute any additional Co_2 into the atmosphere, as the carbon released is the same as the carbon absorbed by the plants as they grow. Using organic fuels is therefore beneficial to both the environment and the atmosphere.

Many potential organic fuels presently pose a waste disposal problem. For example, waste vegetable fats used for cooking require costly disposal. But most of these materials can easily be re-processed to make useful fuels by the process of transesterification.

Biologically derived oils and fats comprise of three fatty acid chains attached to glycerol. Processing detaches the three hydrocarbon chains to make biodiesel, and glycerin. The glycerin can be used to make soaps or fermented to make ethanol, which is re-used to make biodiesel, or it can be burned as a heating fuel.

Biodiesel as an Option for Energy Security in India [1]

India ranks sixth in the world in terms of energy demand accounting for 3.5 per cent of world commercial energy demand in 2001. The energy demand is expected to grow at the rate of 4.8 Per cent per annum. ?? A large part of India's population, mostly in the rural areas, does not have access to it. Hence a programme for the development of energy from raw material, which grows in the rural areas, will go a long way in providing energy security to the rural people. The rise in energy demand in all forms is expected to continue unabated owing to increasing urbanization, standard of living and expanding population stabilization of which not been possible before mid of the current century. The demand of diesel (HSD) is projected to grow from 39.81 million metric tons in 2001-02 to 52.32 million metric tons in 2006-07, at the rate of 5.6 per annum. Our crude oil production as per the Tenth Plan Working Group is estimated to hover around 33-34 million metric tons per annum even though there will be an increase in gas production from 86 million standard cubic metres per day (2002-03) to 103 million standard cubic meters per day in (2006-07). Only with joint venture abroad there is a hope of oil production to increase to 41 million metric tons by (2016-17). The gas production would decline by this period to 73 million standard cubic meters per day. The increasing gap between demand and domestically produced petroleum is a matter of serious concern. Our dependence on imported oil will increase in the foreseeable future. The Working Group has estimated import of crude oil to go up from 85 million metric tons per annum to 147 million metric tons per annum by the end of 2006-07, correspondingly increasing the import bill from $ 13.3 billion to $ 15.7 billion at today's prices. Transport remains the most problematic sector, as no alternative to petroleum-based fuel, has been successful so far. Hence petroleum based fuels especially petroleum diesel (HSD), will continue to dominate the transport sector in the foreseeable future but its consumption can be minimized by implementation of biodiesel programme expeditiously. Targets need to be set up for biodiesel production to achieve blending ratios of 5, 10 and 20 per cent in phased manner.

BIODIESEL REQUIREMENT FOR BLENDING

The estimated biodiesel requirements for blending with petroleum diesel over the period of next five years are given in Table 1:

Table 1. Biodiesel requirement for blending

Year	Diesel demand million tons	Biodiesel requirement for blending million tons		
		@ 5%	@10%	@ 20 %
2001-02	39.81	1.99	3.98	7.6
2002-03	42.15	2.16	4.32	8.9
2003-04	44.51	2.28	4.56	8.6
2004-05	46.97	2.35	4.70	8.4
2005-06	49.56	2.48	4.96	8.9
2007-08	52.33	2.62	5.24	10.48

FEASIBILITY OF PRODUCING BIODIESEL AS A DIESEL SUBSTITUTE

While India is short of petroleum reserve, it has large arable land as well as good climatic conditions (tropical) with adequate rainfall in large parts of the country to account for large biomass production each year. Since the demand of edible oils is higher than their domestic production, there is no possibility of diverting these oils for production of biodiesel. Fortunately, there is a large junk of degraded forest land, unutilized public land, field boundaries and fallow land of farmers where non-edible oil-seeds can be grown. There are many tree species, which bear seeds rich in oil. Of these some promising tree species have been evaluated and it has been found that there are a number of them such as ***Jatropha curcas*** (Ratanjyot) and ***Pongamia Pinnata*** ('Honge' or 'Karanja'), which would be very suitable in Indian conditions. However, ***Jatropha curcas (Ratanjyot)*** has been found most suitable for the purpose. It will use lands which are largely unproductive for the time being and are located in poverty stricken areas and in degraded forests. It can also be planted on farmers' field boundaries and fallow lands. These can also be planted in public lands such as along the railway lines, road sides and beside the irrigation canals.

PROPOSED JATROPHA PLANTATION

Jatropha curcas has been found the most suitable tree species for the reasons summarized below:

- It can be grown as a quick yielding plant even in adverse land situations viz. degraded and barren lands under forest and non-forest use, dry and drought prone areas, marginal lands and as agro forestry crop. It can be planted on fallow lands and along farmer's field boundaries as hedge because it does not grow too tall as well as on vacant lands alongside railways, highways, irrigation canals and unused lands in townships under public / private sector undertakings.

- The seeds of Jatropha are available during the non-rainy season, which facilitates better collection and processing. The cost of plantation is largely incurred in the first year and improved planting material can make a huge difference in yield.
- Raising Jatropha plant and its maintenance creates jobs for the rural poor, particularly the landless, in plantation and primary processing through expellers.
- It has multiple uses and after the extraction of oil from the seeds, the oil cake left behind can be used for bio gas production and is excellent organic manure, the biomass of Jatropha curcas enriches the soil and it can also be put to other uses.
- Retains soil moisture and improves land capability and environment.
- Jatropha adds to the capital stock of the farmers and the community, for sustainable generation of income and employment.

ECONOMICS OF JATROPHA BIODIESEL

In India, it is estimated that the cost of biodiesel produced by transesterification of oil obtained from Jatropha curcas oil seeds shall be approximately same as that of petro-diesel. The seed contains 30 per cent oil, and oil extraction can be 91-92 per cent. 1.05 kg of oil will be required to produce 1 kg of biodiesel, recovery from sale of crude glycerol will be at the rate of Rs. 10 per kg. The price of glycerol is likely to be reduced with processing of large quantities of oil and consequent production of glycerol raising the cost of bio-diesel. However, new applications are likely to be found creating additional demand and stabilizing its price. With volatility in the price of crude, the use of bio-diesel is economically feasible and a strategic option.

ECONOMICS OF BIODIESEL IN THE US

US produces biodiesel from edible oil (mainly Soya oil), the 100 per cent biodiesel costs around $ 1.25 to $2.25 per gallon depending upon purchase volume and the delivery costs and competes with low sulphur diesel oil. However, it is costlier than normal diesel and the B20 blend costs 13 to 22 cents more per gallon than normal diesel. It takes about 7.3 pounds of soya bean oil, which costs about 20 cents/pound, to produce a gallon of biodiesel. Feedstock costs are therefore at least $ 1.5 per gallon of Soya diesel. Under the mustard seed programme, oil can be produced today for approximately 10 cents/pound and the total cost of producing mustard biodiesel is around $ 1 per gallon. The mustard oil, a low value product, contains as much as 90 per cent mono-saturated fatty acids which make it perfect for biodiesel, balancing cold flow issues with No_X emission issues. US is planning to add 5-10 billion gallons of biodiesel through mustard seeds. The mustard oil contains a high value pesticide that helps keep the price of mustard oil low. In India, it is estimated that cost of biodiesel produced by transesterification of oil obtained from Jatropha curcas oil-seeds shall be approximately same as that of petroleum diesel. The byproducts of biodiesel from Jatropha seed are the seed oil cake and glycerol which have good commercial value. The seed oil cake is a very good compost being rich in plant nutrients. It can also yield biogas, which can be used for cooking and its residue as compost. Hence oil cake will fetch good price. Glycerol is produced as a byproduct in the transesterification of oil. These byproducts shall reduce the cost of biodiesel to make it at par with petroleum diesel. The cost components of biodiesel are the price of seed, seed collection and oil extraction, transesterification of oil, transport of seed and oil. As mentioned earlier, cost recovery will be through sale of oil cake and glycerol. The use biodiesel is economically feasible.

Studies on Rapeseed oil methyl ester (RME)

If straight unprocessed vegetable oil is burned, the fatty acids in the oil would start to congeal and harden (coke up) on the inside of the engine as well as in the fuel injectors, eventually leading to big, expensive engine problems. Therefore, it is necessary to process vegetable oil to remove the fatty acids making the oil more similar to diesel compositionally. Biodiesel is an alternative, renewable, clean diesel fuel made from triglycerides (vegetable oils, fats, waste cooking oils), which can be used in neat form, or blended with petroleum diesel for use in compression ignition engines. Biodiesel reduces carbon monoxide, carbon dioxide, sulphur dioxide (one of the main causes of acid rain), hydrocarbons, benzene, and particulate matter. Blending in this way tends to reduce the emissions and smoke levels, though power output will still tend to be lower and separation of the blended constituents could occur in cold weather [3]. No engine problems were reported in large-scale tests with, for example, urban bus fleets running on B20. Fuel economy promised, but it increases nitrogen oxide levels. The biodiesel industry is looking for additives that would reduce nitrogen oxide levels. Kevin et al. [6] compared alternative fuel transit buses price and reported that there were no expected increase for biodiesel blend use because the engine and fuel system were the same as that used for the conventional diesel version. Many scientists reported problems of diesel engine operation with biodiesel. Bag et al. [5] reportessd that extended tests of esters in diesel engines had been encouraging. Like other fuels, combustion of esters is incomplete. The gradual accumulation of esters in the crankcase eventually leads to greatly altered viscosity of the lubricating oil. Bechtold [7] and Perkins et al. [8] reported that biodiesel had higher viscosity and higher pour points compared to typical diesel fuel, which could affect operation in very cold temperatures. Like diesel fuels, pour point additives are effective at decreasing pour point. The above-mentioned problems can be overcome by blending conventional diesel fuel with esters (usually methyl esters) of vegetable oils. The most common ratio is 80 per cent conventional diesel fuel and 20 per cent vegetable oil ester (also termed "B20, indicating the 20 per cent level of biodiesel) like petroleum diesel. No engine modifications are required, and biodiesel maintains the payload capacity and range of diesel. Since modifications are not required, there is no need to change vehicles, spare parts inventories, refuelling stations or skilled mechanics.

Straight biodiesel has a cetane rating significantly higher than typical #2 diesel fuel, slightly lower heating value, slightly higher viscosity, and contains approximately 10 per cent oxygen mass .The lower heating value will cause a small loss in maximum power if the engine fuel system is not recalibrated. Perkins et al. [8] observed performance trends of the diesel and rapeseed methyl ester fuelled engines at 1000 hour of the 1000-hour endurance test and are shown in Figs 1.1 & 1.2. The power and torque curves for the diesel and rapeseed methyl ester fuelled engines are almost similar but slight offset was observed for the rapeseed methyl ester. This offset results in the rapeseed methyl ester fuelled engine reaching their peak power output at a lower engine rpm. Another notable difference is that the rapeseed methyl ester fuelled engine produced a flatter power curve than the diesel-fuelled engine. Babu et al. [9] has worked both on the neat vegetable oils and their esters. He has studied the power and torque curves of the engine with the usage of above said oils, which are depicted below:

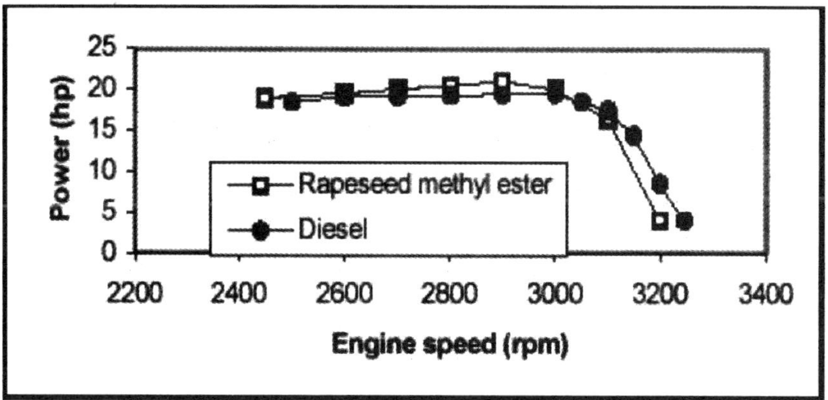

Fig. 1.1. Power output for rapeseed methyl ester and diesel fuel

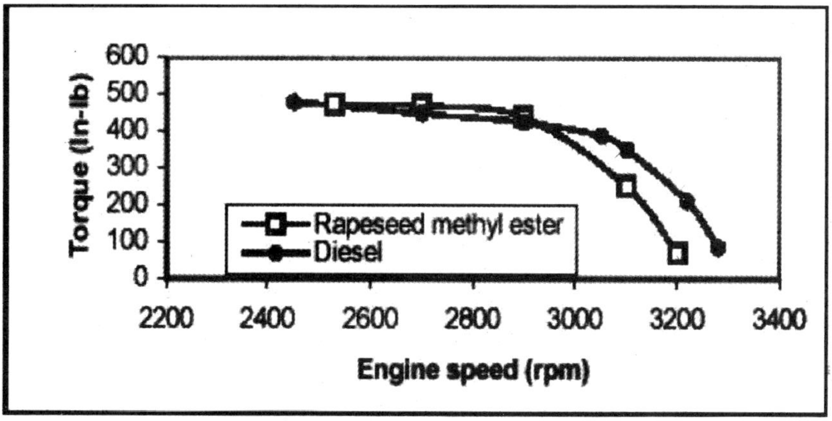

Fig. 1.2.Torque for rapeseed methyl ester and diesel fuel.

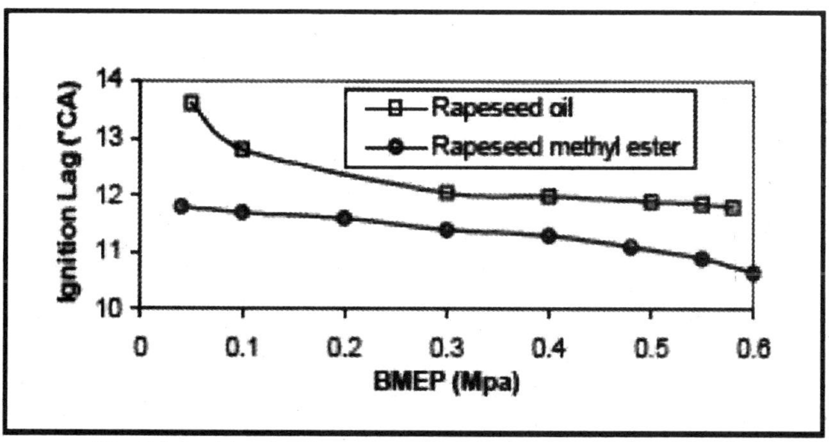

Fig. 1.3. Ignition lag for rapeseed methyl ester and diesel fuel

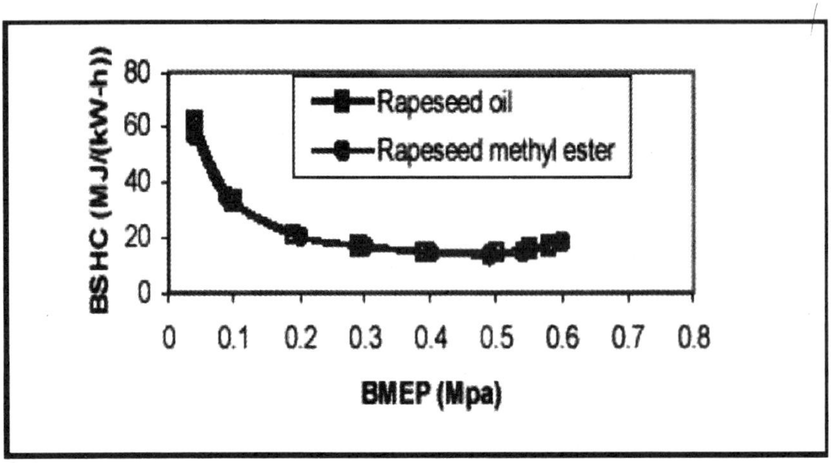

Fig. 1.4. BSHC for rapeseed methyl ester and diesel fuel

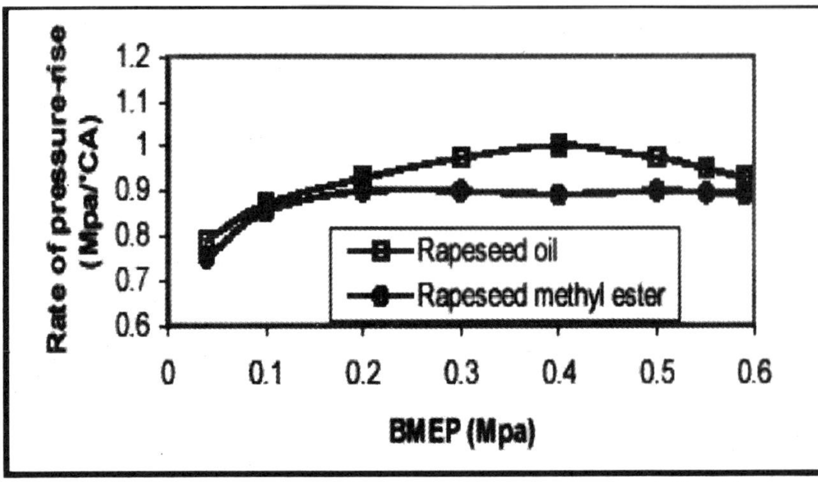

Fig. 1.5. Rate of pressure-rise for rapeseed methyl ester and diesel fuel

Murayama et al. [2] compared the engine performance with the rapeseed methyl ester with pure rapeseed oil. The difference in the engine performance between the two fuels is very small, except for the ignition lag, which is smaller for the methyl ester (Figs. 1.3 & 1.4). However, the major difference between the two fuels is the much smaller carbon deposits with methyl ester when compared with pure rapeseed oil. It can be seen from Fig. 1.5 that the engine noise increases linearly with an increase in the amount of premixed combustion and the maximum rate of pressure rise, regardless of the kind of fuel. Therefore, the maximum rate of pressure rise and the amount of premixed combustion appear to be critical factors for controlling the engine noise.

Senatore et al. in their paper [15] works on the RME inspired by the results obtained [10-14] on bio fuel which has good overall behaviour, with performance and emission levels comparable to diesel fuel.

These figures present the torque and specific consumption respectively under the above test conditions. For the various fuel properties, whenever the comparison is made for the same equivalence ratio value, there is observed a marked difference in the curves (Figs. 1.6a, 1.7a and 1.8a.), which stresses how the performance of biodiesel is clearly lower (about 20-25%) than that obtained from diesel fuel, when the same quantity of air and fuel is introduced into the cylinder. As expected, this difference tends to cancel itself out if one refers to the relative equivalence ratio 'φ' (Figs.1.6b, 1.7b and 1.8b).

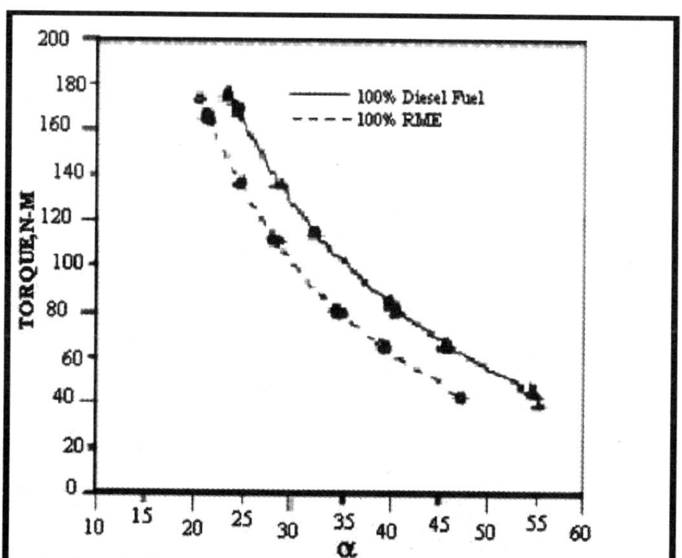

Fig. 1.6a. Torque Vs α

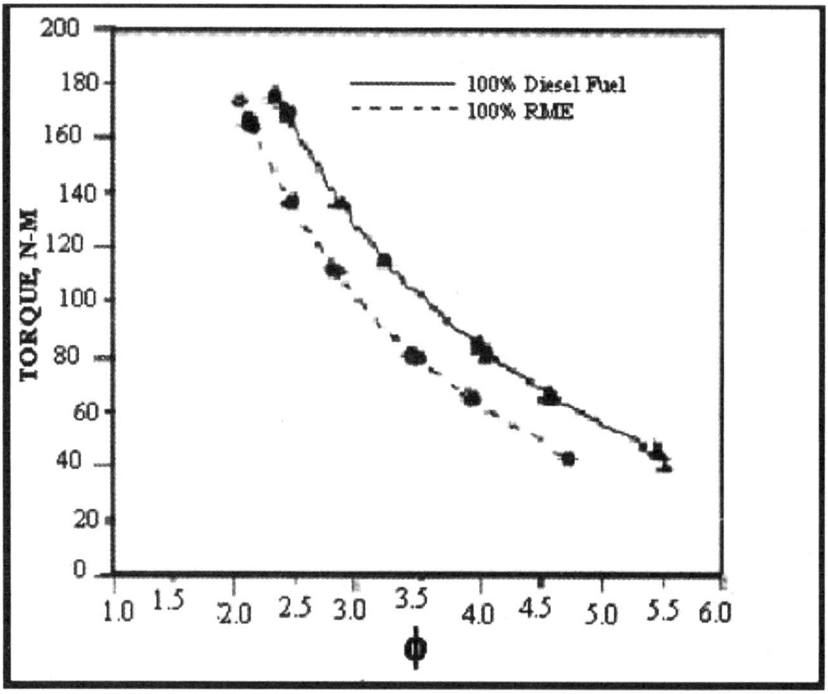

Fig. 1.6b. Torque Vs φ.

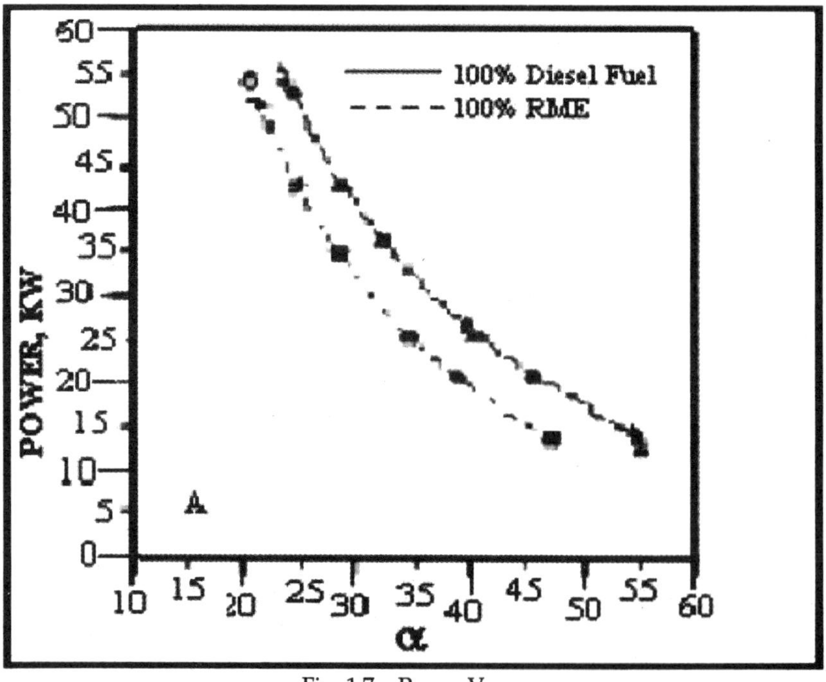

Fig. 1.7a. Power Vs α

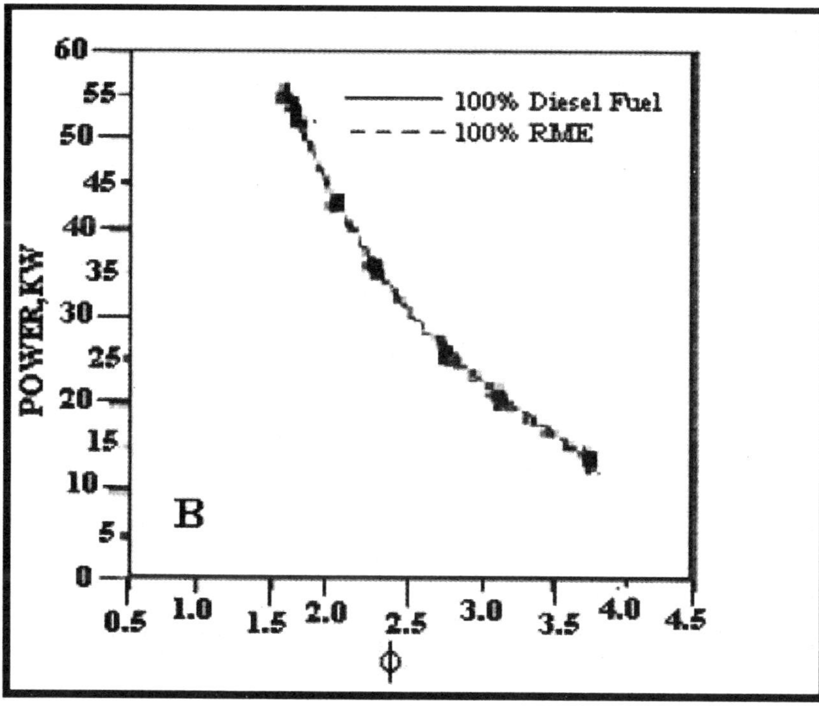

Fig. 1.7b. Power Vs φ

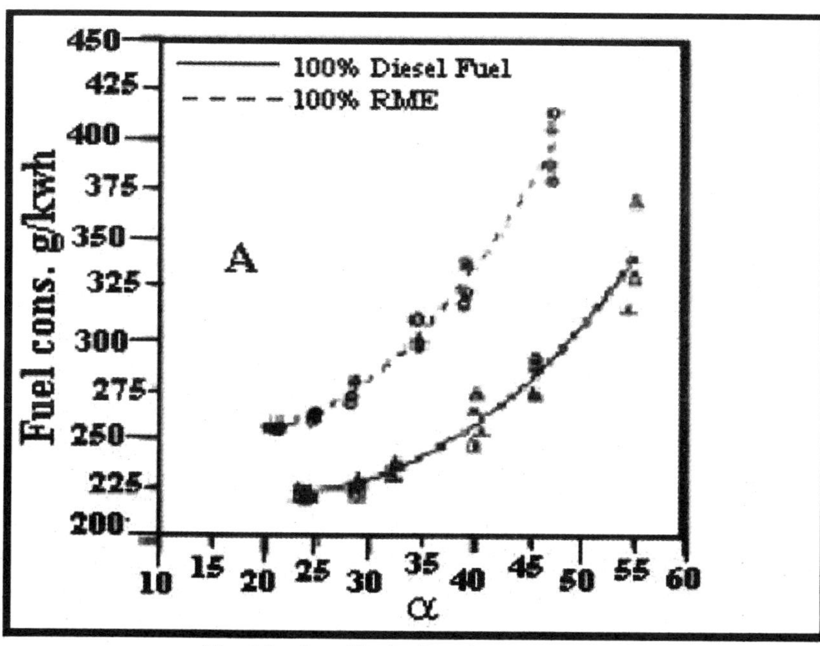

Fig. 1.8a. Specific fuel consumption Vs α

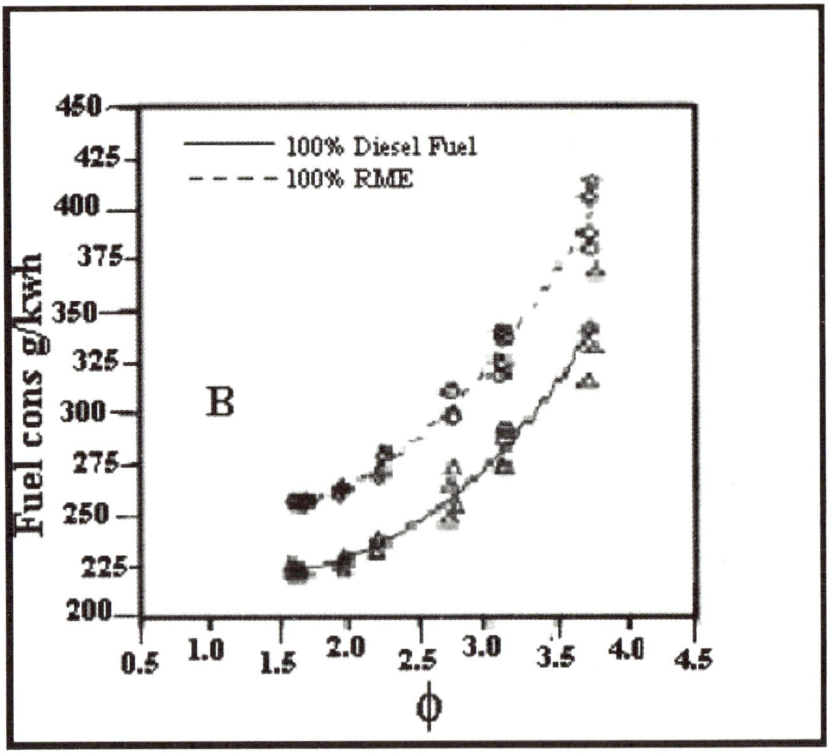

Fig. 1.8b. Specific fuel consumption Vs φ

Figures 1.9-1.11 shows the net heat release rate diagrams for three different operating conditions (φ = 3.75, φ = 3.15 and φ = 1.61) for pure diesel fuel and pure methyl ester, respectively. These figures point out that the heat release rate initially follows a downward trend, corresponding to the end of compression stroke, which suddenly changes slope at combustion starting. By analyzing these diagrams, one can observe that when the engine is fuelled with biodiesel, the process starts in advance in all operating conditions, a feature, which becomes more evident as the load increases. This determines a similar trend in the mean temperature variation rate of gases in the cylinder, as shown in Figs.1.12-1.14. Therefore, depending on the fuel used, the maximum temperature increase rate is found at somewhat different engine crank angle positions. In particular, compared with the operating conditions tested using diesel fuel, when biodiesel is used the temperature variation rate peaks at a position closer to piston top dead centre.

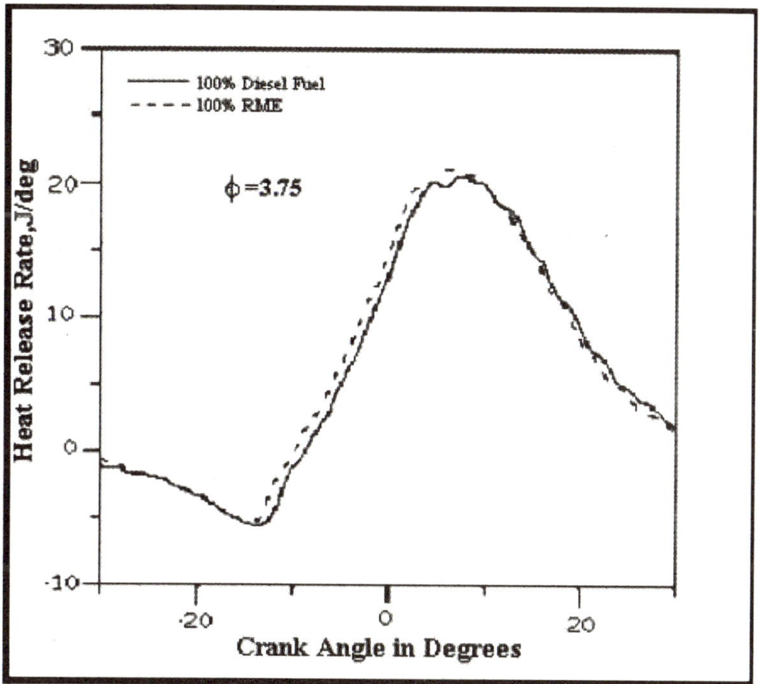

Fig. 1.9. Heat release rate Vs crank angle in degrees.

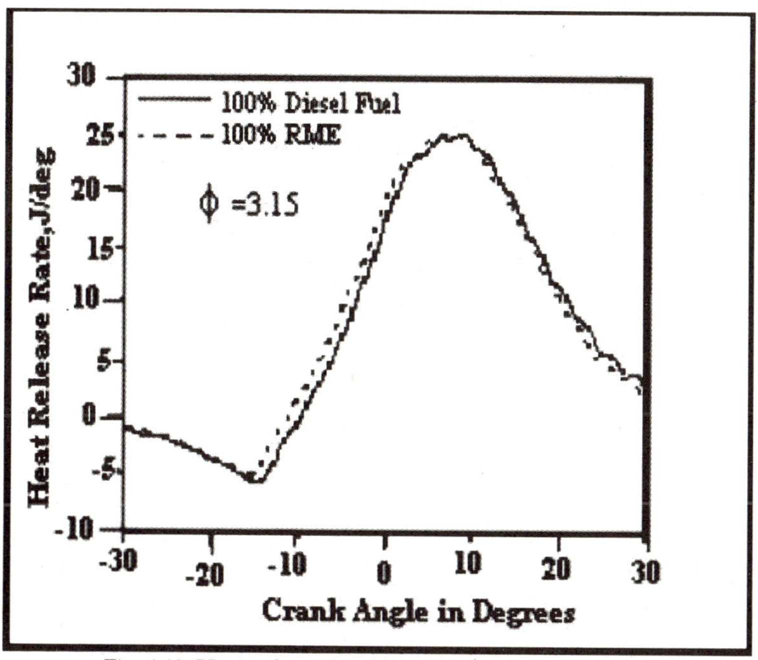

Fig. 1.10. Heat release rate Vs crank angle in degrees.

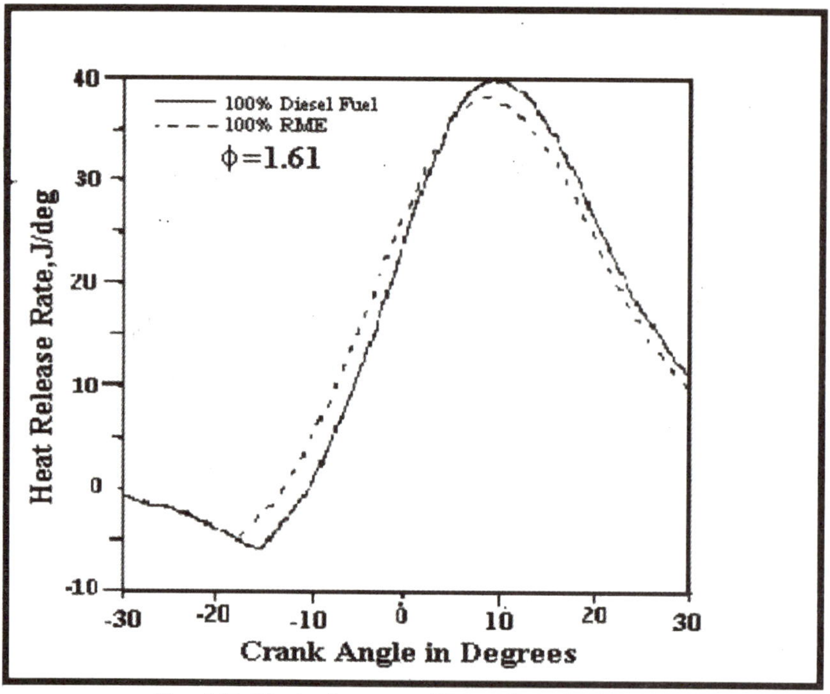

Fig . 1.11. Heat release rate Vs crank angle degree

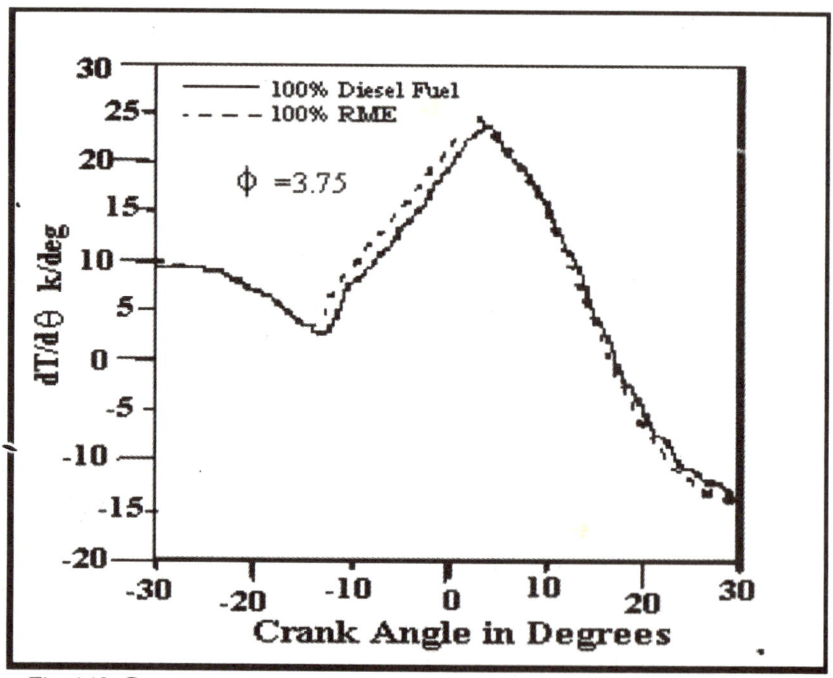

Fig. 1.12. Gases mean temperature variation rate Vs crank angle in degrees.

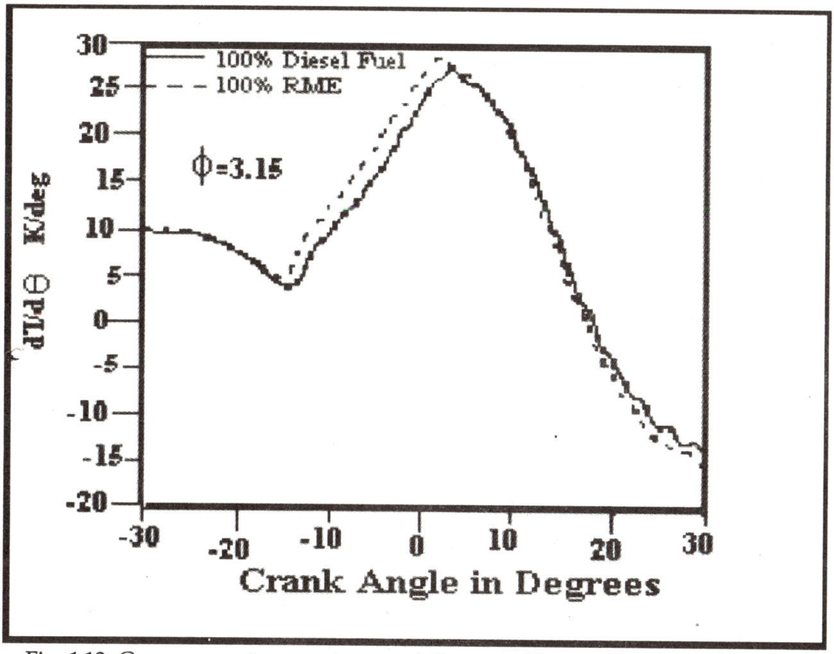

Fig. 1.13. Gases mean temperature variation rate Vs crank angle in degrees.

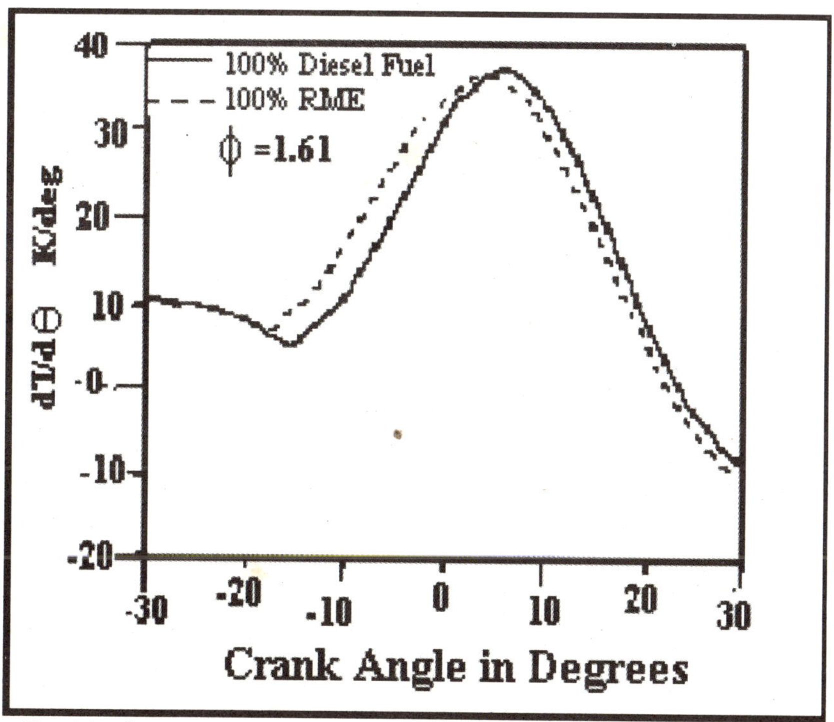

Fig. 1.14. Gases mean temperature variation rate Vs crank angle in degrees.

Desantes et al. [16] evaluate the potential of rapeseed oil methyl ester (RME) to improve the combustion process in a high-speed direct injection (HSDI) diesel engine equipped with high-pressure common-rail injection system. The study, based on the comparison of three different fuels (standard gas-oil, RME and 30% RME/gas-oil mixture), takes into account the main aspects that control diesel combustion, from the injection rate characteristics to the spray behaviour characterized using an optical pressurized chamber.

This global study of the whole injection-combustion process identifies some causes of the decrease in pollutant emissions observed when the engine operates with RME.

The use of fuels derived from agricultural products like rapeseed oil methyl ester (RME) is currently receiving considerable attention due to its economic and ecological interest [18] [19]. Several authors have evaluated RME behaviour in D .I. and I.D.I. diesel engines [20] [23] [24] [25], and previous results are variable, particularly regarding pollutant emissions. Experiments have involved the study of mechanical stress problems of the injection system and oil contamination [20] [21] [22]. In the cases where detailed studies of the combustion process with RME have been carried out, the analysis was purely chemical [17]. Very few studies have incorporated an analysis of the influence of the RME characteristics on the physical mechanisms that control the injection-combustion process in a diesel engine.

Studies on Sunflower oil methyl ester (SME)

These are the observations made by Dulger et al.[26] by his experimentation on sunflower methyl ester. The variation of full load torque with engine speed for biodiesel and diesel fuel is given in Fig. 1.15 For diesel fuel, the maximum torque is 152 Nm at 2250 rpm while for SME, the maximum torque is 150 Nm at 2250 rpm. Maximum torque is obtained at same rpm for both fuels. At 4500 rpm, diesel gives a torque of 117 Nm, biodiesel 106 Nm. Overall, diesel fuel yields 0-10 per cent higher torque values than SME. Figure 1.16 shows full load engine power variation with engine speed. A maximum power of 55 kW is obtained at 4500 rpm for diesel, maximum power for SME is 50 kW at 4500 rpm. Below 2500 rpm, power outputs for both fuels are almost same. Above 2500 rpm, diesel produces 0-10 per cent higher power than SME. The reason why SME produces lower torque and power output than diesel fuel is attributed to fuel pumping problems of SME due to its higher density and viscosity compared to diesel, lower heating value of SME and inappropriate injection timing for SME. Figure 1.17 shows specific fuel consumption. Minimum specific fuel consumption for diesel is 280 g/kW-h at 2000 rpm, while for SME 283 g/kW-h at 2000 rpm. At 4500 rpm, diesel consumes 326 g/kW-h and SME 340 g/kW-h. Over the whole rpm range, fuel consumption for SME is 2-5 per cent higher. The reason for this is believed to be lower heating value of SME than diesel fuel which leads to higher fuel consumption per unit power produced. No performance penalties [27], [28], [29], [30], contamination of fuel line elements and lube oil and carbon deposits are found in the combustion chamber due to high viscosity [31].

This study shows that SME is an alternative to diesel fuel. According to the tests, the torque, power and specific fuel consumption for SME operation are within the same levels as when operating with pure diesel fuel. Although the results of the tests carried out on the test bench seem to be very encouraging, more tests with SME should be carried out to cover all operating conditions, not only full load conditions. Moreover, modifications on engine design and operation parameters such as injection timing, injection pressure and fuel heating should be tested and optimized for SME operation.

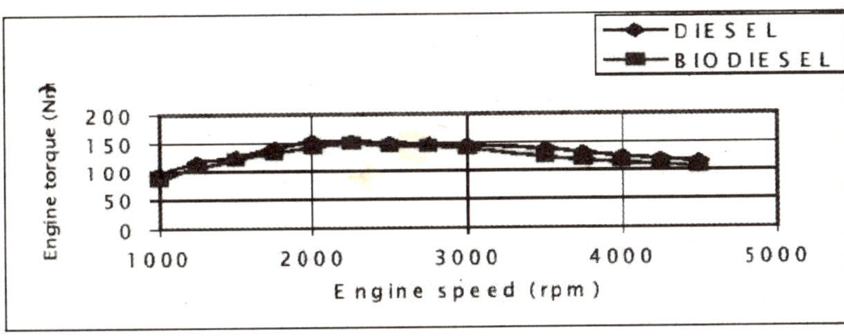

Fig. 1.15. Engine torque variation with engine speed

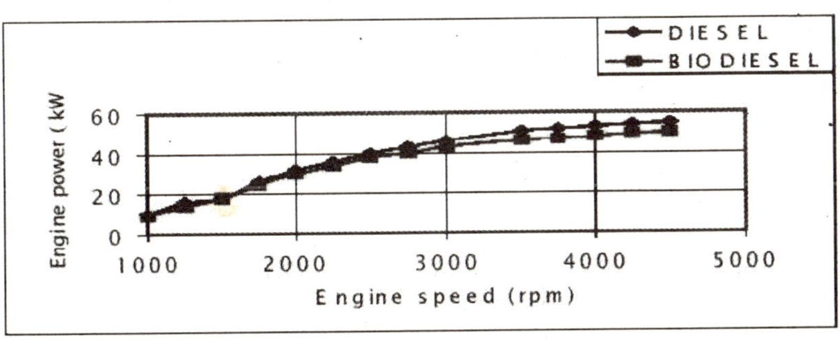

Fig. 1.16. Engine power variation with engine speed

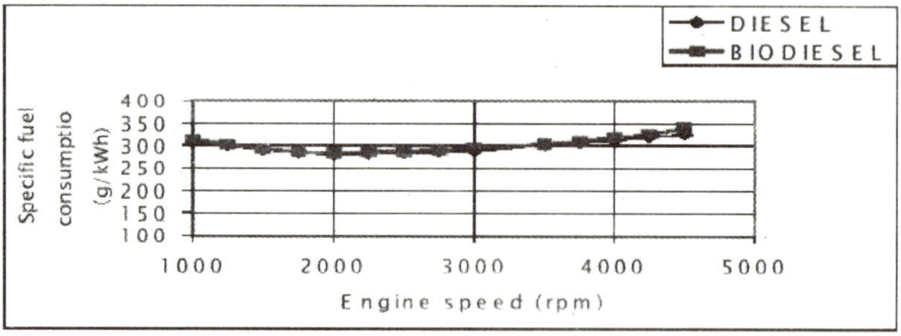

Fig. 1.17. Specific fuel consumption with engine speed

Studies on Palm oil methyl ester (POME) [70]

Tests were carried out by Palm Oil Research Institute of Malaysia (PORIM) with ordinary diesel and POME. A brief description of the conditions and properties is given below:

a) Baseline Ordinary diesel fuel
b) POME 2535 POME at 25° C and intake air at 35° C
c) POME 9535 POME at 95° C and intake air at 35° C
d) POME 5035 POME at 50° C and intake air at 35° C
e) POME 5050 POME at 50° C and intake air at 50° C
f) POME 8585 POME at 85° C and intake air at 85° C
g) POME 2550 POME at 25° C and intake air at 50° C
h) POME 2575 POME at 25° C and intake air at 75° C

ENGINE TRIALS

Brake Power

Referring to Fig.1.18, the maximum brake power is achieved at around 2000 r/min, conforming very well with the manufacture's specifications. It can be observed that the engine produced almost identical performance on OD and POME fuel under different preheated conditions. It can be seen that as the fuel temperature is increased there is a tendency to power improvement, especially while the engine is operating below 3500 r/min, which was best observed in the case of POME9535 and POME5035. This is most probably due to the reduction of fuel viscosity as temperature rises; once a POME is heated up to 55°C its viscosity approaches the OD level (refer to Table 2). The reduction in viscosity leads to fuel droplet refinement, which in turn improves the fuel spray, atomization, fuel evaporation and injection characteristics of POME fuel. In addition, the ignition delay time was reduced and the cetane rating was also increased, thus approaching the minimum specification for diesel fuel (13). This leads to 'diesel-like' brake power output. However, as the fuel is heated above 100°C the fuel viscosity becomes too low to effectively perform its lubricating function on the fuel pump; which deteriorates fuel pump delivery efficiency and results in lower power output (at 95°C POME viscosity is about 2.2 cSt). Combustion was more nearly 'hypergolic' with preheating since the time for each fuel element to be oxidized was significantly reduced. Pursuant to the phenomenon of hypergolic combustion, thermal dissociation can be achieved by preheating the fuel. Upon injection of preheated fuel into the combustion chamber, owing to the presence of a few chemically active fuel radicals, the fuel is ignited and consumed more rapidly than fuel injected at conventional temperatures. The actual phenomenon of hypergolic combustion can be realized only if the preheated fuel is subject to a high supply pressure with the aid of a high-temperature fuel injection system. This condition is still under investigation. For the case of POME 2550, the hot intake air activated the air-fuel mixture in the combustion chamber, therefore improving combustion efficiency, which in turn leads to better power output.

Torque

The plot of torque Vs engine speed is shown in Fig.1.19 indicating a similar trend as in Fig. 1.18. Thus the above arguments are used to explain the observation.

Specific fuel consumption

A plot of specific fuel consumption against engine speed is shown in Fig.1.20. It is shown that POME at normal and preheated forms exhibits slightly higher specific fuel consumption than OD fuel. The differences in fuel consumption reflect the differences in relative fuel density, calorific value of fuel, fuel quality and quality of fuel in terms of the cetane index. Referring to Table 2, the POME has a higher specific density (around 5 per cent higher than OD), lower specific combustion enthalpy and lower heat release rate than OD fuel. The higher specific gravity of POME compared with OD fuel indicates that a 15 per cent lower calorific value by mass leads only to a 10 per cent reduction in volume. Since diesel injection equipment metres by volume, the maximum fuel energy delivery without modification is effectively reduced by 10 per cent. In other words, an extra 10 per cent of POME fuel is required to produce the same level of energy output as OD fuel. In addition, once POME fuel is preheated to above 55°C, its viscosity is reduced significantly; this will increase the fuel flow rate, which in turn will increase the specific fuel consumption. The preheated fuel shows similar fuel characteristics, but with a proportional reduction in viscosity with the increase in temperature, which is clearly shown in Table 3. POME 5035 shows a similar level of fuel consumption with baseline OD fuel, as their viscosity levels are equivalent. The trends demonstrated in Fig1.20 also indicate that an increment in fuel temperature has more adverse effect on specific fuel consumption for POME than the intake air temperature.

Brake Thermal Efficiency

Figure 1.21 shows the brake thermal efficiency against engine speed. It is obvious that the baseline OD fuel performs the best in terms of brake thermal efficient throughout the engine operation range. This is mainly due to the relatively higher power output and lower fuel consumption when using baseline OD fuel. However, for the preheated POME cases, even though they generated slightly higher power output than the baseline OD system, the adverse increment in fuel consumption is a serious penalty to downgrade their brake thermal efficiencies.

Table: 2. I Properties of O.D. and Palm Oil Methyl Ester (POME)

Properties	POME	OD
Specific density (g/cm^3)	0.875	0.832
Kinematic viscosity at 40° C	4.17	3.60
Cetane number	50-52	53
Calorific value (KJ/ Kg)	41300	46800

Viscosities of POME at various temperatures are shown in Table 3.

Table 3. Viscosities of POME and Temperature (ASTMD445)

S.No.	Temperature in ^{0}C	Viscosity
(1)	29.5	8.05
(2)	70	4.40
(3)	55	2.93
(4)	100	2.06
(5)	120	1.65

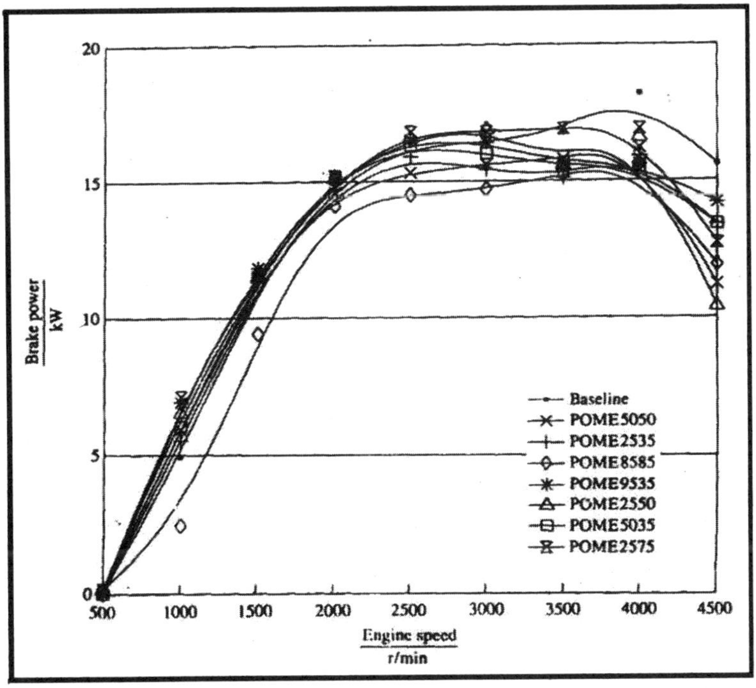

Fig. 1.18. Brake power Vs engine speed

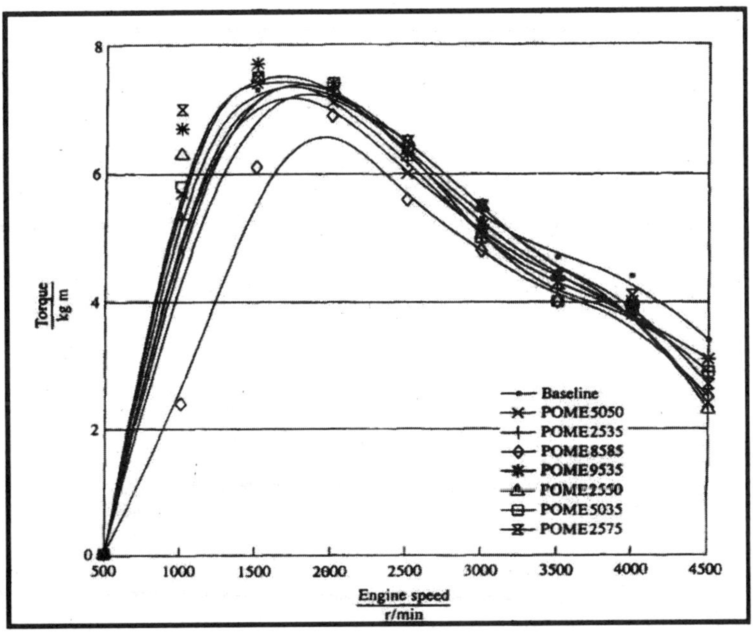

Fig. 1.19. Torque Vs engine speed

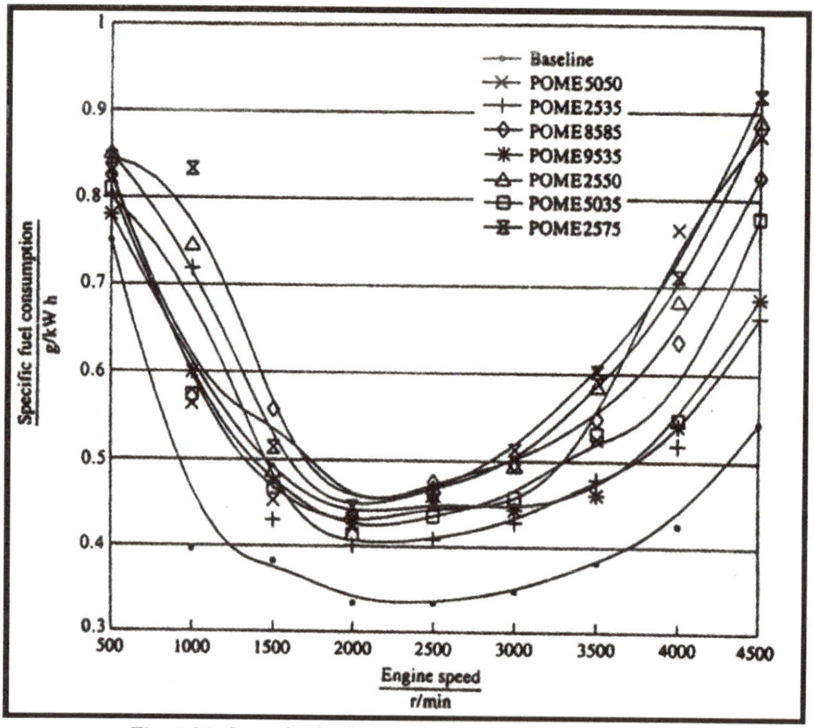

Fig. 1.20. Specific fuel consumption Vs engine speed

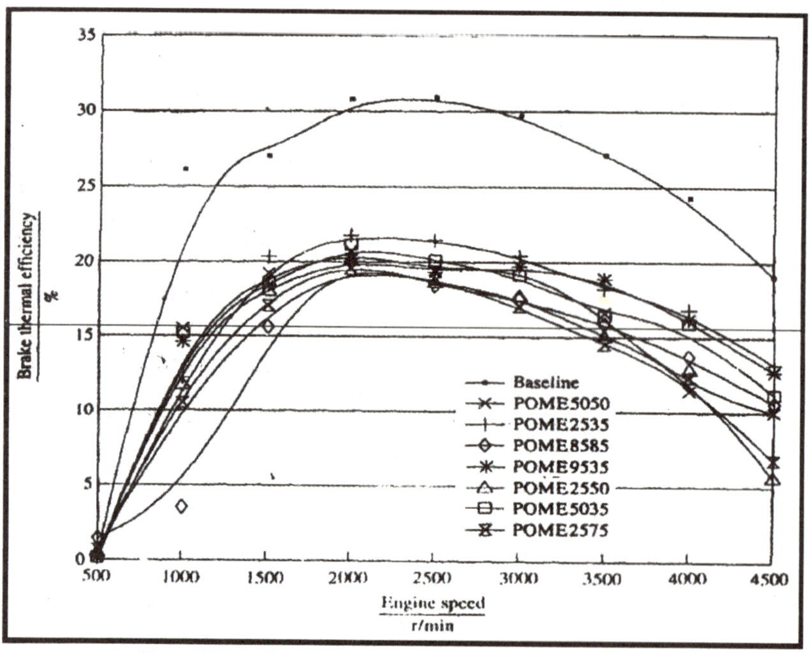

Fig. 1.21. Brake thermal efficiency Vs engine speed

Studies on Jatropha oil Methyl Ester (JME)

Uncertainties concerning adequate and stable supplies of petroleum fuels have renewed interest in vegetable oils as diesel engine fuels [32]. Non-edible oils are quite promising fuels for agricultural applications. Vegetable oils have similar properties to diesel and can be used to run compression ignition engines with few or no modifications [33]. Investigations have been carried out on a variety of vegetable oils, *e.g.*, Jatropha oil, Karanja oil, rice bran oil, rapeseed oil, for use in diesel engines [34]. Diesel engines with vegetable oils as fuels produce the same power output but with reduced thermal efficiency and increased emissions [35]. Besides this, vegetable oils lead to problems of gum formation and smoke emission. Most of the problems of vegetable oils can be overcome by esterifying them.

Kumar et al. [36], in their experimental work, have evaluated and compared the performance, emissions and combustion characteristics of a single-cylinder, water-cooled, direct-injection diesel engine running on Jatropha oil and its methyl ester with diesel operation. With Jatropha oil, the maximum brake thermal efficiency was 25.6 per cent compared to 29.4 per cent with diesel. The thermal efficiency was 27.8 per cent with the methyl ester of Jatropha oil. Smoke and particulate levels were higher with Jatropha oil than diesel. The methyl ester of Jatropha oil was better than the pure oil but it was still inferior to diesel. Hydrocarbon and carbon monoxide emissions were slightly higher with both Jatropha oil and its methyl ester than with diesel whereas NO levels were lower than with diesel. Ignition delay and combustion duration were longer and peak heat release rates were lower with Jatropha oil and its methyl ester than with diesel. Again the methyl ester of Jatropha oil was better. On the whole, it is concluded that the methyl ester of Jatropha oil could be used unmodified in diesel engines with no major detonation in performance. The long-term effects of this fuel have to be evaluated.

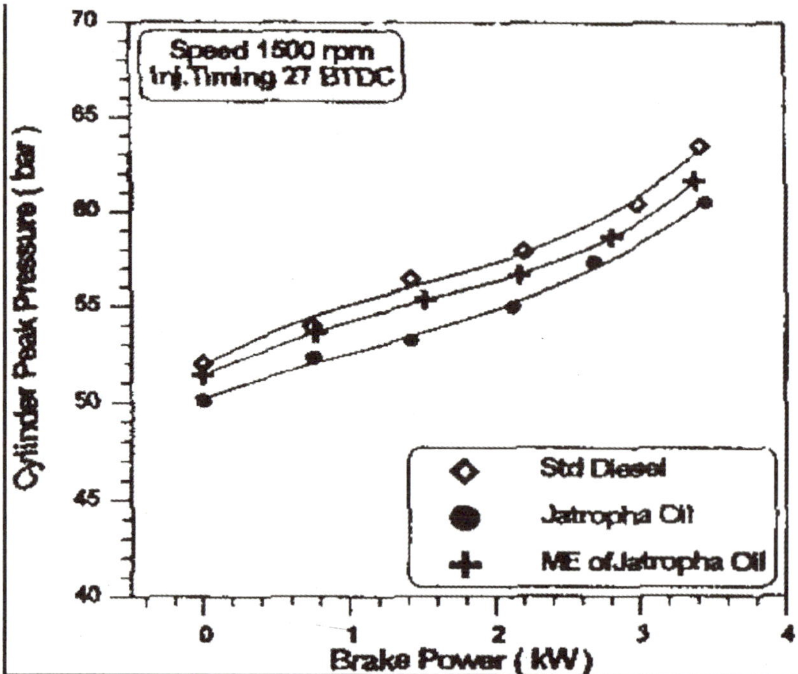

Fig . 1.22. Variation of peak pressure with brake power

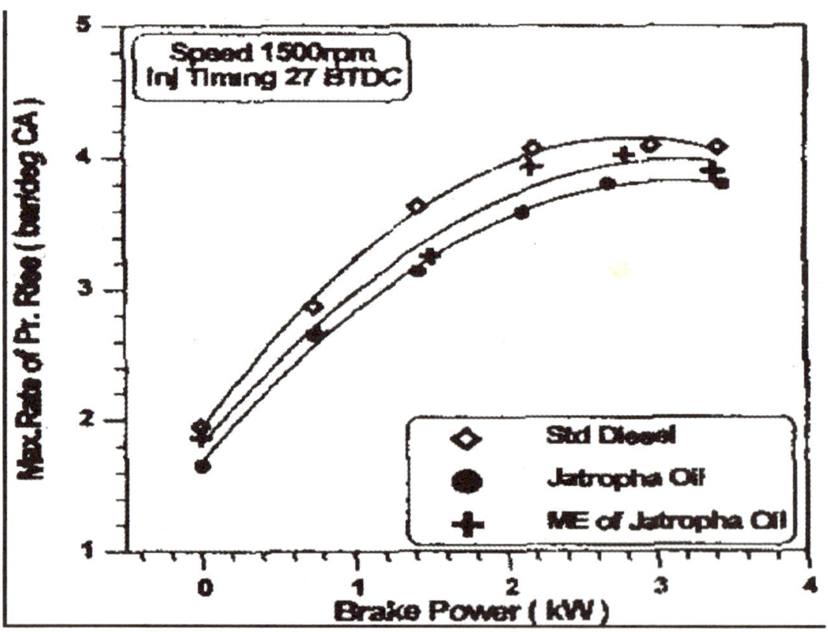

Fig. 1.23 .Variation of maximum rate of pressure rise with brake power

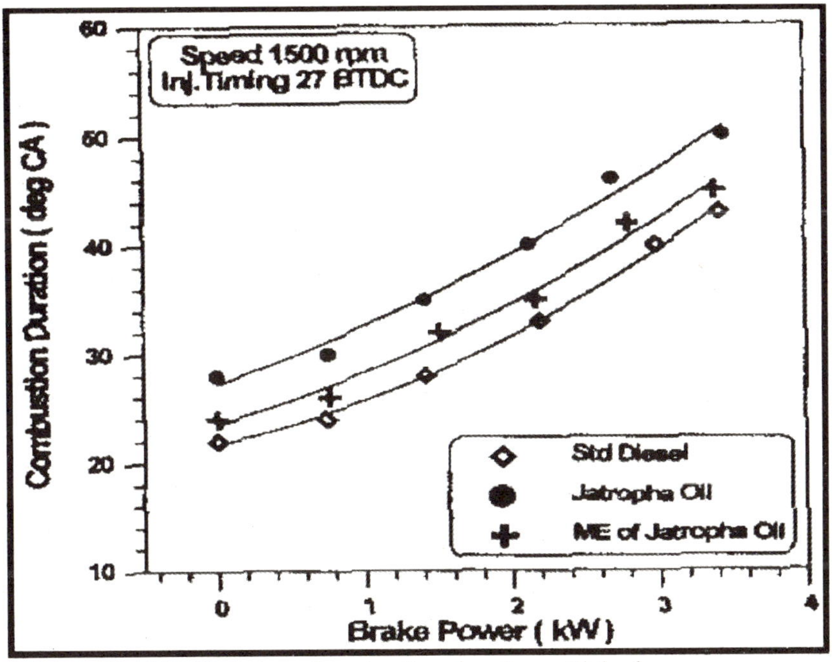

Fig .1.24 .Variation of combustion duration with brake power

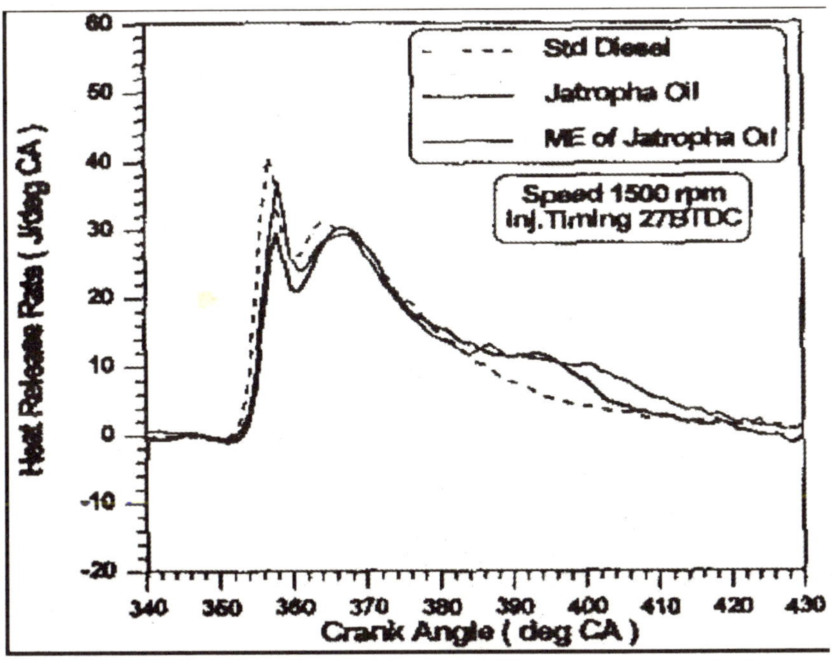

Fig. 1.25. Variation of heat release rate at Maximum power output

A single-cylinder diesel engine was operated on Jatropha oil, methyl ester of Jatropha oil and diesel. The following are the conclusions based on experimental results:

- The operation of the engine is smooth on Jatropha oil and methyl ester of Jatropha oil with acceptable performance. The use of Jatropha oil results in a slightly lower thermal efficiency than with diesel. With the methyl ester of Jatropha oil the brake thermal efficiency is similar to diesel values. Maximum brake thermal efficiencies are 25.6, 27.8 and 29.4 per cent for Jatropha oil, its methyl ester and diesel, respectively.
- Exhaust gas temperature is higher with the Jatropha oil and methyl ester of Jatropha oil than with diesel due to slow combustion.
- Hydrocarbon emission is greater for the Jatropha oil than the diesel, 150 pp m for Jatropha oil and 120 pp m for diesel at maximum output. However, it is only 130 ppm with the methyl ester of Jatropha oil. Similar trends are seen in the case of CO emission.
- The maximum smoke level with Jatropha oil is 4.4 BSU, whereas it is only 3.8 BSU with its ester and 3.6 BSU for diesel.
- Particulate emissions are lower with methyl ester of Jatropha oil than pure Jatropha oil. Trends are similar to smoke emissions.
- Ignition delay and combustion duration are longer with both Jatropha oil and methyl ester of Jatropha oil than with diesel.
- Lower rates of heat release are found with Jatropha oil and methyl ester of Jatropha oil than with diesel during the premixed combustion phase.

On the whole, it is concluded that the methyl ester of Jatropha oil can be good alternative fuel for diesel engine, however, the problem of sluggish combustion and high smoke levels needs attention. The long-term effects of these fuels have to be evaluated.

The performance of the engine and combustion parameters of the engine with diesel and methyl ester of Jatropha oil are presented and discussed below.

The variation of specific fuel consumption and brake thermal efficiency with power output for the methyl ester of Jatropha oil and diesel are shown in Figures 1.26 and 1.27. The specific fuel consumption is higher i.e., the thermal efficiency is lower with the ester of Jatropha oil. This is probably due to the low volatility, slightly higher viscosity and high density of the methyl ester of Jatropha oil, which affects mixture formation of the fuel and leads to slow combustion. The maximum thermal efficiency of the Jatropha ester is about 26.5 per cent, whereas it is 29.3 per cent with diesel. However, there is no drop in maximum power with the Jatropha ester.

Exhaust gas temperature increases when the load rises and it is higher for Jatropha ester than diesel, particularly at higher loads. Due to incomplete combustion of the injected fuel and part of the combustion extending into the exhaust stroke, there is a slight increase in exhaust gas temperature with Jatropha ester compared to diesel oil operation. The maximum temperature of exhaust gas at peak load is 492°C for the ester and 474°C for diesel. With 100 per cent methyl ester of Jatropha oil' operation smoke emission is increased at higher loads. This may be due to the higher viscosity and density of the ester that leads poor vaporization and slow combustion of the injected fuel. When the highly viscous Jatropha ester is injected, the atomization of fuel is poor leading to larger droplets and less air entrainment resulting in inefficient combustion. This leads to higher smoke emission with Jatropha ester. However, smoke is less at lower loads.

The variation of peak pressure and the rate of pressure rise for various loads are shown in Figures 1.28 and 1.29. The peak pressure depends on the amount of fuel taking part in the uncontrolled combustion phase, which is governed by the delay period and the spray envelope of the injected fuel. There is a little difference between the peak pressures of the methyl ester of Jatropha oil and diesel. At high loads the peak pressure is slightly higher than diesel with the Jatropha ester. As seen in Figure 1.29, there is a big difference in the maximum rate of pressure rise between the two fuels. The maximum rate of pressure rise of the ester of Jatropha oil is generally higher at high power outputs probably due to the dominating effects of the premixed phase of combustion.

INFERENCE

A single cylinder compression ignition engine was operated successfully using methyl ester of Jatropha oil as the sole fuel. The following conclusions are made based on the experimental results:

- Engine works smoothly on methyl ester of Jatropha oil with performance comparable to diesel operation.
- Methyl ester of Jatropha oil results in a slightly reduced thermal efficiency as compared to that of diesel.
- The exhaust gas temperature is increased with the methyl ester of Jatropha oil as compared to diesel operation at peak output.
- Hydrocarbon emission is low with the methyl ester of Jatropha oil.

- There is no significant difference in smoke emissions when the methyl ester of Jatropha oil is used.
- CO emission is increased at higher loads with the methyl ester of Jatropha oil. On the whole, it is concluded that the methyl ester of Jatropha oil will be a good alternative fuel for the diesel engine for agricultural applications.

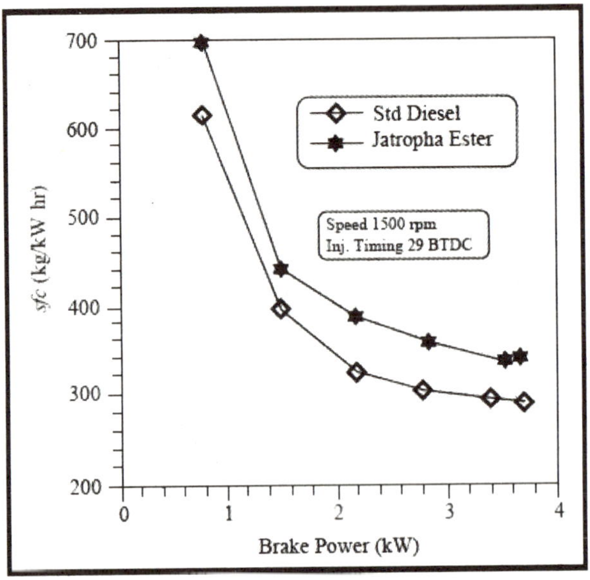

Fig. 1.26. Variation of *sfc* with brake power

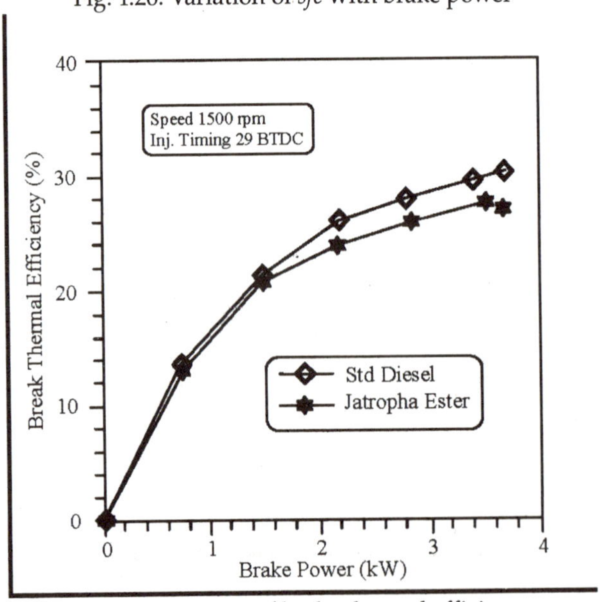

Fig ,1.27. Variation of brake thermal efficiency.

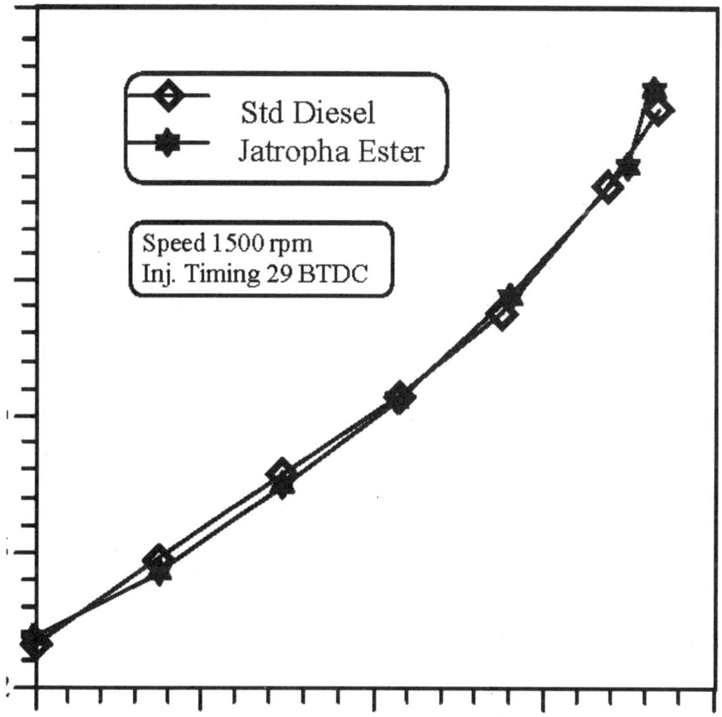

Fig 1.28 .Variation of peak pressure with brake power

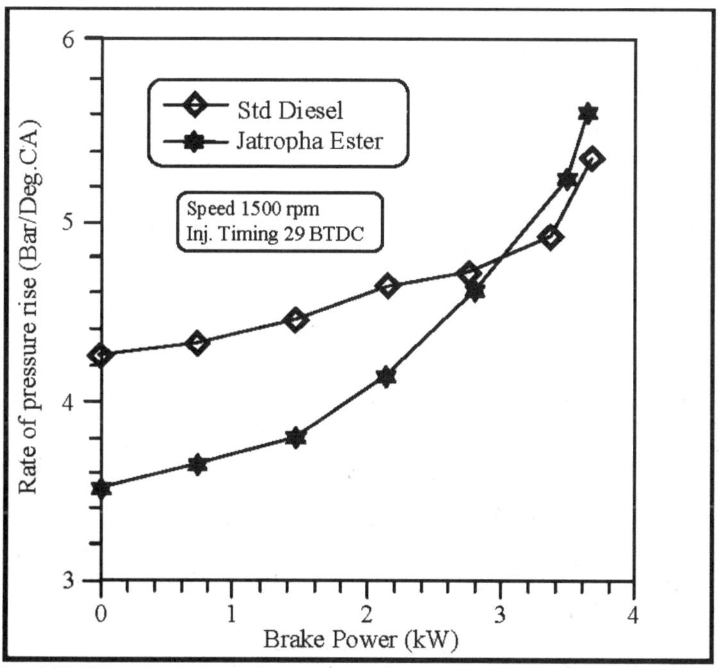

Fig. 1.29. Variation of maximum rate of pressure rise with brake power

Studies on methyl tallowate biodiesel

Ali et al. [37] have conducted experiments on four fuels produced by blending biodiesel (methyl tallowate) and #2 diesel fuel. Engine in-cylinder pressure data were collected at various engine speeds and used to evaluate the peak pressure, indicated mean effective pressure (*imep*), rate of change of pressure, rate of heat release, mass fraction of fuel burned, and charge temperature with respect to crank angle. Peak cylinder pressures for each fuel blend at all engine speeds were lower than peak pressure for #2 diesel fuel, the *imep* values for all fuel blends were less than that of #2 diesel fuel. The differences in *imep* values correlate with the differences on power output of the engine. The maximum rates of pressure rise for all fuel blends are less than that of #2 diesel fuel. The rate of heat release decreased with increase in engine speed as well as with the amount of methyl tallowate in the fuel blend. Peak rate of heat release for all fuel blends is less than that of #2 diesel fuel. When methyl tallowate was blended with #2 diesel fuel, the shift in the location of peak heat release was slightly away from top dead centre (TDC). Ignition delay slightly increased when methyl tallowate was blended with #2 diesel fuel at peak torque conditions but at peak power the type of fuel blends did not affect ignition delay. On the other hand, burn duration slightly decreased with increase in methyl tallowate in the fuel blend at peak torque condition and there was no significant effect on burn duration at peak power condition. The charge temperature decreased with increase in methyl tallowate content of the fuel blends. A reduction in charge temperature can help reduce NO_x emissions. It was concluded that the fuel blends used in this study would have no detrimental long-term effects on engine performance, wear, and knock.

Interest in cleaner burning fuel is growing worldwide and reduction in exhaust emissions from the internal combustion engine is gaining importance from the past few years. It is widely recognized that alternative diesel fuels produced from vegetable oils and animal fats can reduce the exhaust emissions from compression ignition engines without significantly affecting engine performance. Ali et al. [38] reported no significant difference in power output with different blends of diesel fuel, methyl tallowate, methyl soyate, and ethanol. Further, CO and NO_x emissions were not affected by the blends used, but HC emissions were significantly lower with a 80:13:7 per cent (v/v) blend of diesel: methyl tallowate and ethanol, the recommended blend for minimum emissions as compared with #2 diesel fuel.

Pollutant emissions reduction from diesel engines requires detailed knowledge of the combustion process. However, the complex nature of the combustion process in a diesel engine makes it difficult to understand the events occurring in the combustion chamber which determine the emissions of exhaust gases including CO, HC, NO_x and smoke. Most CI engines are designed to operate on diesel fuel and, therefore, perform best while operated on that fuel. During the engine design and optimization process, an engine manufacturer performs in-cylinder pressure measurement to determine cylinder pressure, rate of change in pressure, estimated rate of heat release, mass-burned fraction, and charge temperature. High-speed data acquisition systems are used by performance development engineers and are found to be of practical value. Algorithms and techniques that provide an accurate representation of heat transfer and a means of very accurately determining top dead centre (TDC) have been developed. Engine cycle analysis (ECA) and fuel injection analysis (FIA) software can be used to obtain steady state engine performance characteristics (Gill.A.P. [41]). The combustion process in a diesel engine is usually considered to occur in four phases according to heat release rate (Barbella et al. [39]). Those phases are the ignition delay period, premixed burning phase, diffusion burning, after diffusion burning phase and oxidation phase. These phases are used to follow the transformation of fuel in the combustion cycle.

Several studies have been reported on the effects of fuel and engine parameters on diesel exhaust emissions. Kittelson et al. [42] conducted in-cylinder and exhaust soot mass measurement on a single-cylinder conversion of a four-cylinder, direct injection (DI) diesel engine using a sampling system, which allowed dumping, diluting, quenching, and collection of the entire contents of the cylinder on a time scale of about 1 ms. They observed that soot concentration, heat release, and fuel injection data were related to one another. Maximum soot mass was observed shortly after top dead centre (ATDC), which reached a peak between 15 and 30^0 crank angle (CA) ATDC. After reaching its peak value, soot concentration decreased with increasing CA and approached exhaust levels by 40 to 60^0 CA ATDC. There was a longer delay between the start of combustion and the start of soot formation for high equivalence ratio, which was due to a slightly longer premixed burning phase at high load. The increase in length of the premixed burning period was much smaller than the increase in the formation lag time. They concluded that oxygen availability late in the cycle was a critical factor in determining exhaust soot concentrations.

Shundoh et al. [44] observed that low (1,000 rpm) and high (2,000 rpm) engine speeds had no effect on heat release rate because the injection rates were the same, and heat release closely followed the injection rate in type C combustion. The gas temperature in the case of the low speed condition was higher than that at high speed and was the reason for higher NOx and lower smoke levels at low speeds.

Barbella et al. [39] studied the formation and oxidation of soot, light and heavy hydrocarbons, CO, CO_2, and NO_x during the combustion cycle of a DI diesel engine. They observed that the concentrations of heavy hydrocarbons decreased during the early stages of the combustion cycle. Maximum soot formation occurred ATDC after that the soot formation decreased slowly at 40^0 CA ATDC.

Niehaus et al. [43] observed that diesel fuel produced more premixed burning than thermally cracked soyabean (TCSBO) at brake mean effective pressures (*bmep*) of 100 and 300 KPa.

Czerwinski et al. [40] used a rapeseed oil, ethanol, and diesel fuel blend and compared the heat release curves with diesel fuel. He observed that the addition of ethanol caused longer ignition lag at all operation conditions. At higher and full loads, the combustion speeds were high with strong premixed phases. The addition of rapeseed oil gave a little lower combustion speed and lower combustion temperature as compared to diesel fuel. The inflammation lag with rapeseed oil was slightly shorter and the combustion duration was approximately equal to diesel fuel. The blend of diesel, rapeseed oil, and ethanol had lower heat values, which diminished the power output at full load as well as the available power during the transient operating conditions.

The overall objective of this project is to perform in-cylinder pressure measurements on an engine to determine cylinder pressure, rate of pressure change, rate of heat release, mass-burned fraction of fuel, and charge temperature curves on fuels produced by blending #2 diesel fuel and biodiesel (methyl tallowate) in a Cummins N14-410 diesel engine. It is expected that this study would help establish the fuel burning characteristics needed to control exhaust emissions and engine coking.

COMBUSTION ANALYSIS

Pressure vs Crank Angle

This relationship gives a gross indication of engine knock, the location of peak pressure, and the value of the peak pressure. Peak cylinder pressures for each fuel blend at all engine speeds were lower than the peak pressure with #2 diesel fuel. Representative graphs showing pressure with respect to crank angle for peak torque and rated engine speed are shown in Figures 1.30 and 1.31, respectively.

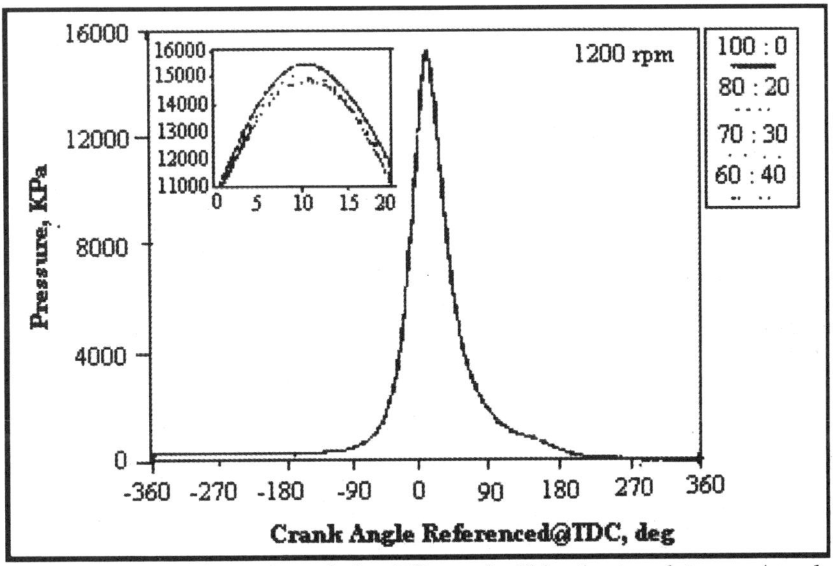

Fig. 1.30. Pressure Vs crank angle for different fuel blends at peak torque (graph in box shows the peak pressure values)

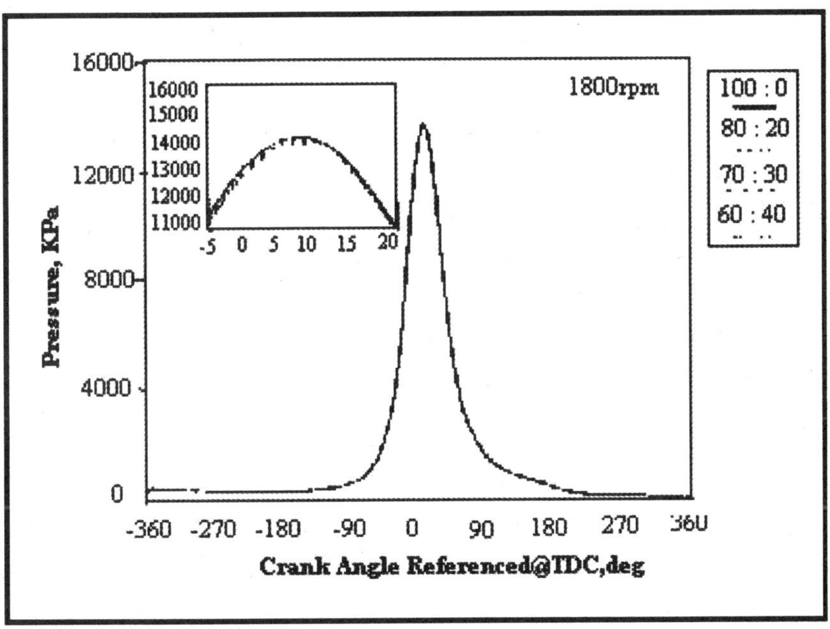

Fig. 1.31. Pressure Vs crank angle for different fuel blends at rated engine speed (graph in box shows peak pressure values)

Similar trends are observed at all other engine speeds and loads. Looking at the magnitude of the peak pressures for the blends as compared to #2 diesel fuel, it was concluded that since the peak pressures were less than that for #2 diesel fuel, there should have been no effect on engine durability and there should not be any problem related to knock, combustion, or partial burn with engine performance.

Indicated Mean Effective Pressure

Indicated mean effective pressures is defined as the indicated average constant pressure exerted on the piston during the expansion stroke, which will produce the same amount of work as the actual pressure during the compression and expansion strokes. Since the *imep* is related to the power output of the engine, differences in the *imep* can be compared with differences in power output at a given engine speed.

At 1200 rpm engine speed, the *imep* values for the 80:20, 70:30, and 60:40 per cent (v/v) blends of diesel: methyl tallowate were within 2.5 per cent of that for #2 diesel fuel. These results were consistent with the peak pressure results with the same blends. At rated speed of 1800 rpm, the *imep* values for the above blends were all within 1 per cent of that for #2 diesel fuel. Also the indicated mean effective pressures for all fuel blends and engine speeds were less than the *imep* for #2 diesel fuel. The differences in *imep* values corresponded with differences in power output of the engine.

Derivative of Pressure with Respect to Crank Angle

This analysis, which is simply the derivative of the pressure shown in Figure 1.30, indicates how rapidly pressure changes and helps identify potentially damaging combustion conditions. After correlating the observed value of the maximum rate of pressure rise with the engine hardware, an engine manufacturer can establish a limiting maximum value, which will ensure acceptable engine life.

The representative graphs showing the development of change in pressure with CA for engine speed at peak torque and rated engine speed are shown in Figures 1.32 and 1.33, respectively. The points in the combustion strokes at which the derivatives are equal to zero in Figures 1.32 and 1.33 correspond to the points of maximum pressure indicated in Figures 1.30 and 1.31, respectively.

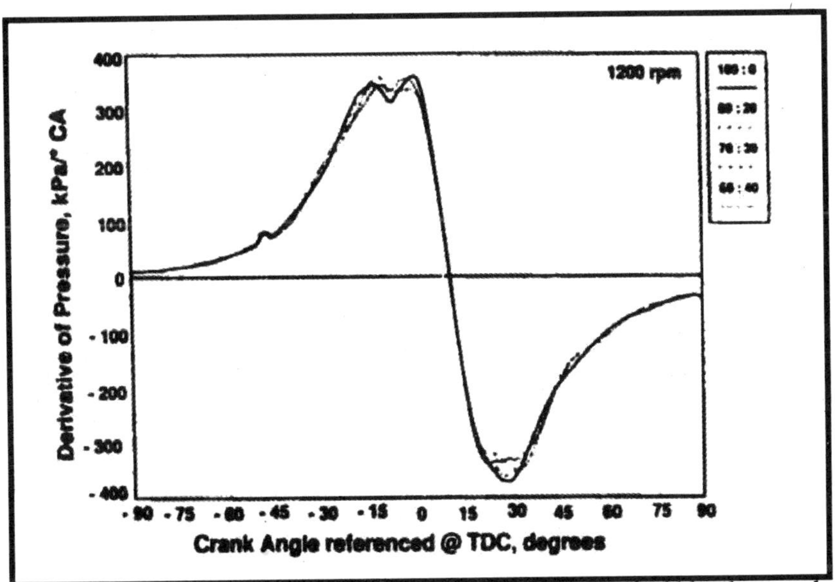

Fig. 1.32. Derivative of pressure Vs crank angle for different fuel blends at peak torque

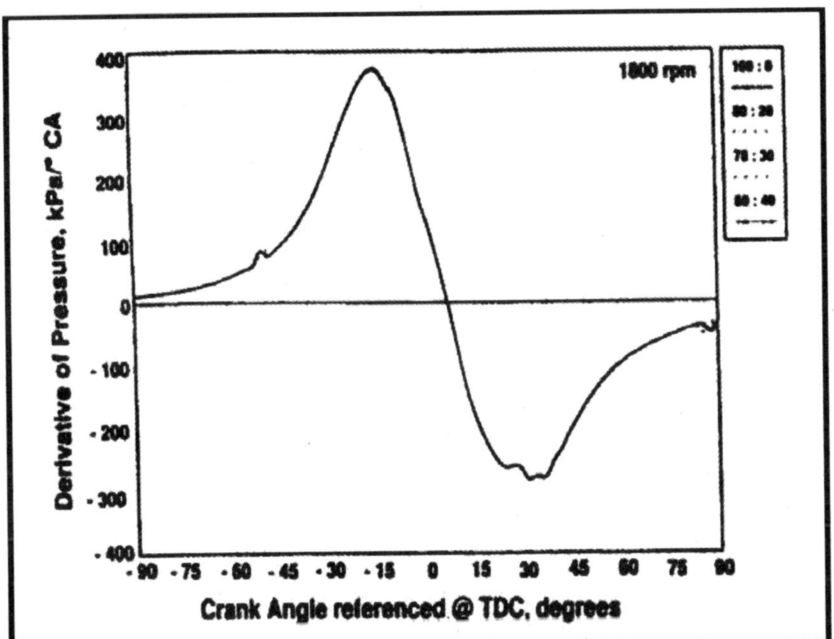

Fig. 1.33. Derivative of pressure Vs crank angle for different fuel blends at rated engine speed

The results showed that in-cylinder pressure change was maximum at an engine speed of 1100 rpm and decreased as engine speed was increased to 1200 rpm and then increased as engine speed was further increased to 1600 rpm, the speed at which maximum power was produced. Further increase in engine speed again decreased the peak value of pressure change. The location of maximum pressure change with CA also shifted from 0.0 to 0.2^0 CA, ATDC TO 13 to 13.8^0 CA BTDC (before top dead centre). The shapes of all pressure change curves were the same as that for #2 diesel fuel. #2 diesel fuel has a peak change of pressure of 361, 385 and 374 KPa at engine speed of 1200, 1600, and 1800 rpm, respectively. The locations shift from 0.2^0 CA ATDC to 13.6^0 CA BTDC and 13.8^0 CA BTDC at the above engine speeds, respectively.

At 1200 rpm engine speed, peak values of change in pressure with crank angle for the 80:20, 70:30, and 60:40 per cent (v/v) blends of diesel: methyl tallowate are within 2 per cent of that for #2 diesel fuel. The location of these points were 0.2^0 CA ATDC, 11.2^0 CA BTDC and 2.4^0 CA BTDC for the respective fuel blends. At an engine speed of 1600 rpm, where peak power is produced, the peak value of change in pressure was maximum within the operating range of 1200 to 1800 rpm. The peak values of change in pressure for these fuel blends were within 1.5 per cent of that for #2 diesel fuel and their locations were more or less the same as that for #2 diesel fuel. At rated engine speed of 1800 rpm, the peak values of change in pressure with CA for all fuel blends were within 1.5 per cent of that of #2 diesel fuel with their locations shifted to 14.6, 14.4, and 14.0^0 CA BTDC respectively, as compared to 13.8^0 CA BTDC, for #2 diesel fuel.

Looking at the performance of the engine in terms of peak value of change in pressure with CA, it can be concluded that there were minor differences in the peak values. In most cases, the peak value of change in pressure was less than that for #2 diesel fuel. These small differences in the rate of change in pressure should have no long-term effect on engine performance, wear, and knock.

Rate of Heat Release Vs Crank Angle

This analysis shows the estimated rate of heat release during the combustion process. The results provided a quantified assessment of combustion rate and the means to diagnose combustion problems. The analysis was based on pressure and volume measurements. Therefore, some assumptions were made to calculate the rate of heat release. The first assumption was that the trapped charge remained in a uniform single zone of constant composition from intake valve closing to exhaust valve opening. Actually, large temperature gradients existed in the charge during combustion and the chemical composition of the unburned gases was different from the burned gases. The second assumption was that leakage and heat transfer to the wall was negligible. The third assumption was that the charge mixture behaved as an ideal gas. Based on these assumptions, the rate of heat release with respect to CA and location of peak heat release for blends of diesel and methyl tallowate at different engine speeds were calculated and it was observed that peak rate of heat release decreased as engine speed increased from 1100 to 1900 rpm. The location of peak rate of heat release was delayed as the engine speed increased. Furthermore, as the diesel fuel content of the blended fuel was reduced, the peak rate of heat release was also reduced. The shapes of peak rate of heat release curves for all fuel blends at all engine speeds were similar to that of #2 diesel fuel. #2 diesel fuel had a peak rate of heat release of 0.287 KJ/^{0} CA at the engine's peak torque producing speed of 1200 rpm and 0.250 KJ/^{0} CA at the engine's rated speed of 1800 rpm. The trends of peak rates of heat release with CA for all fuel blends at engine speeds of 1200 and 1800 rpm are shown in figures 1.34 and 1.35, respectively.

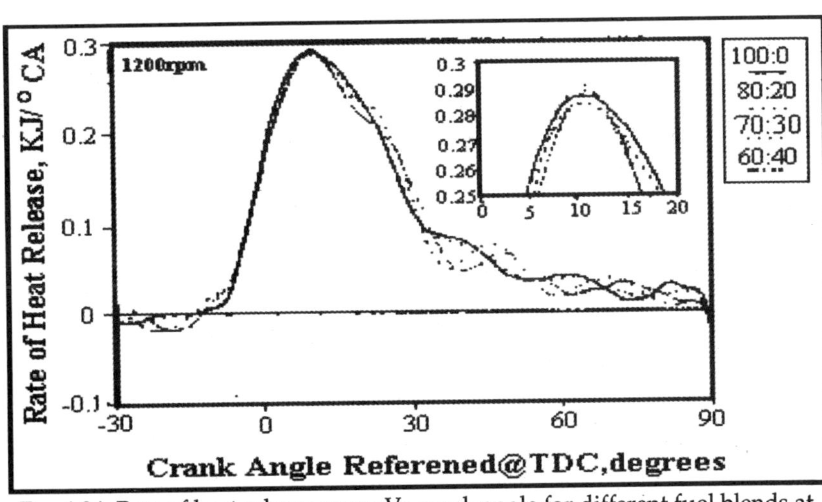

Fig. 1.34. Rate of heat release curve Vs crank angle for different fuel blends at peak torque (graph in the box shows the peak rate of heat release)

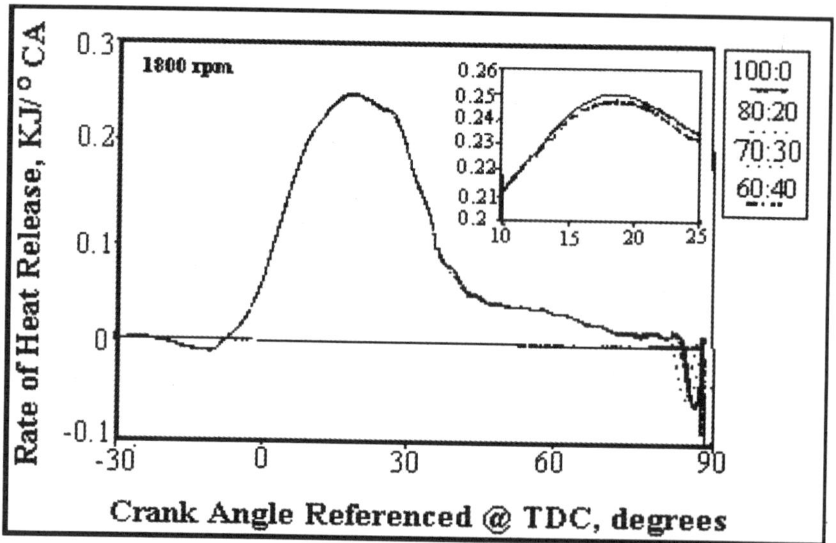

Fig. 1.35. Rate of heat release curve Vs crank angle for different fuel blends at rated engine speed (graph in the box shows the peak rate of heat release)

At 1200 rpm engine speed, peak rates of heat release for 80:20, 70:30 and 60:40 per cent (v/v) blends of diesel: methyl tallowate are within 2 per cent of that for #2 diesel fuel. The locations of peak rates of heat release are within 10.6 and 11.2^0 CA after top dead centre (ATDC) for the respective fuel blends as compared to 10.6^0 CA ATDC for #2 diesel fuel. At rated engine speed of 1800 rpm, the peak rate of heat release for the 80:20, 70:30, and 60:40 per cent (v/v) blends of diesel: methyl tallowate were within 1.2 per cent of that of #2 diesel fuel. Locations of peak rates of heat release were within 0.4^0 CA ATDC of that of #2 diesel fuel. A similar trend was observed at all other engine speeds. In general, as the amount of methyl tallowate was increased, there were reductions in the peak rates of heat release as compared to #2 diesel fuel with the exception of the 70:30 and 60:40 per cent (v/v) blends of diesel: methyl tallowate, in which cases there are slight increases at 1200 rpm. Reductions in the peak rate of heat release are less than that of #2 diesel fuel (Ali et al. [55]). To understand the process of heat release in detail, one must know the mass-burned fraction of the fuel with respect to CA to determine the ignition delay and burn duration.

Vegetable oils have comparable energy density, cetane number, heat of vaporization, and stoichiometric air/fuel ratio with mineral diesel fuel. The large molecular sizes of the component triglycerides result in the oils having higher viscosity compared with that of mineral diesel fuel. The viscosity of liquid fuels affects the flow properties of the fuel, such as spray atomization, consequent vaporization, and air/fuel mixing. The problem of viscosity has an adverse effect on the combustion of vegetable oils in the existing diesel engines. Besides, some problems crop up in the associated fuel pump and injector system. An acceptable alternative fuel for engines has to fulfill the environmental and energy security needs without sacrificing operating performance [55], [56].

Methanol and ethanol are two abundantly available alternative fuels, which possess the potential to be produced from biomass sources. These fuels can be successfully used as diesel engine fuels by preparing biodiesel. Transesterification process utilizes methanol or ethanol and vegetable oils as the process inputs. This indirect route of utilizing alcohol as a diesel engine fuel is definitely a superior route as the toxic emissions containing aldehydes are drastically reduced. The problem of corrosion of various engine parts utilizing alcohol as fuel is also solved by the way of transesterification [57].

Biodiesel is a chemically modified alternative fuel for diesel engines, derived from vegetable oil, fatty acids, and animal fat. In its simplest form, the carbon cycle of vegetable oils consists of the fixation of carbon and the release of oxygen by plants through the process of photosynthesis and then combining of oxygen and carbon to form CO_2 through the processes of combustion or respiration. It is appropriate to mention here that the CO_2 released by petroleum diesel was fixed in the atmosphere during the formative years of the earth, whereas the CO_2 released by biodiesel gets continuously fixed by plants and may be recycled by the next generation of crops. The carbon cycle time for fixation of CO_2 and its release after combustion of biodiesel is quite small as compared (few years) to the cycle time of petroleum oils (few million years). It is well known that petroleum refiners are now facing new sulphur and aromatic compound specifications. Since biodiesel is a fuel made up of esters derived from oils and fats from renewable biological sources, it has been reported to emit far less regulated pollutants than petroleum diesel fuel [58],[60].

The process of utilizing biodiesel in the IC engines for transport as well as other applications is gaining momentum recently. IEA (International Energy Agency) has recognized biodiesel as an alternative fuel for the transportation sector. The European Commission proposed a 12 per cent market share for biofuels by the year 2020. Kaltschmitt et al. [57] conducted a study, which shows that beanery carriers offer some clear ecological advantages over fossil fuels such as conserving fossil energy resources or reducing the greenhouse effect [63].

Knothe et al [4] investigated the influence of fatty acid structure on the performance as a diesel fuel. Pischinger et al [59] found monoesters to be technically suitable in terms of good miscibility with diesel oil, almost similar volumetric heat content, adequate viscosity and cetane number [65]. The high carbon residue indicated by the Conrad son value and high viscosity is due to the large molecular mass and chemical structure. The high carbon residue is likely to lead to heavy smoke emission from an engine [66].

In Austria, biodiesel from rapeseed oil is commercialized. The first industrial plant for biodiesel production with a capacity of 10,000 tons per annum in the world went into operation in Austria in 1991. RME is produced in co-operative and industrial-scale plants. An interesting fact is that biodiesel is produced from a mix of RME, SME, and FME (used frying oil methyl ester) in Austria. Tax benefits in Austria and Germany encourage the use of 100 per cent biodiesel fuel in ecologically sensitive areas and agricultural and mountainous region [64], [65]. In Ireland, two pilot projects on the use of biodiesel (RME) in commercial vehicles are reported. In the first project, pure biodiesel was used in buses, trucks, and in pleasure cruisers. The RME used was obtained through oil extraction, followed by esterification on a small-scale tractor-mounted esterification unit. The second project involved a comparative analysis of the behaviour of a number of vehicles operated on RME and SME, imported from UK and Italy, respectively. Vehicle testing started in mid-1996. A principal disadvantage of rapeseed oil as a source of biodiesel in Ireland was its economic feasibility, as it is up to 0.25 ECU/litre (about 6.6 ECU/GJ) more expensive than its fossil-based equivalent. In Denmark, two companies produced non-food rapeseed oil, on a commercial basis, for export. At present, there is no commercial production of biodiesel in Denmark. During a test, four city buses were made to run on biodiesel for 3 months. The results were promising but not satisfactory from environmental point of view. The Danish National Transport Plan "Trafik 2000" mentions the use of biofuels as one of the five major instruments to reduce CO_2 emissions from the transport sector. However, there is no commitment for large-scale use within the government [61],[62],[66]. In the United States, fuel tax rebate or other governmental aid or regulations could propel biodiesel into a high volume, lower cost production track that enables it to compete head-on-head with diesel fuel for a variety of applications [59].

American Society for Testing and Materials (ASTM) has prescribed certain tests and their limits for diesel fuel to be used in CI engines. For any alternative fuel suitable for long-term engine operation without engine modifications, it should be in conformity or within close range to these ASTM limits [67].

Studies on Linseed oil methyl ester (LOME)

A host of plant and forest resources are available in the world from which different vegetable oils can be produced and formulated for use in diesel engines. The present work has been carried with such an objective, where non-edible oil has been appropriately modified by way of esterification and subsequently used for running the diesel engine. Linseed oil was selected for the present investigation, as it is available in large quantities and non-edible in nature. Linseed oil is obtained from the dried ripe seed of the flex plant, ***Linum usitaissimum***, grown in the temperate areas across the world. Its viscosity is lower than most of other vegetable oils. It has high linoleic acid content. Linoleic acid is a straight chain molecule of 18 carbon atoms [C_{17} H_{29} COOH] with three double bonds at 9-10, 12-13, and 15-16 carbons. Its high degree of unsaturation is responsible for the drying properties of the oil. Linseed oil has lower heating value (LHV) of 39.75 MJ/kg (dry).

This reduction in frictional losses is also reflected by the decreasing trend followed by unaccounted losses, which can be noticed from the heat balance sheets for various concentrations. The energy saved by decrease in frictional horsepower makes additional contributions towards useful energy, cooling losses, and exhaust losses. This reflects in increased thermal efficiency, cooling losses and exhaust losses. It is evident from the graphs of exhaust temperature vs. *bmep* (Fig. 1.36 and Fig. 1.39) that the exhaust gas temperature also increases along with thermal efficiency as a function of blend concentration initially. Smoke opacity for biodiesel blends (Fig.1.37 and Fig.1.38) is also noticed to be generally lower than that of diesel oil. It also repeats the trend followed by thermal efficiency curves. There are two main reasons: the higher thermal efficiency means better and complete combustion and lesser amount of unburned hydrocarbons in the engine exhaust thus improving smoke opacity values.

The molecule of biodiesel, i.e., linseed oil methyl ester (LOME), contains some amount of oxygen that takes part in combustion and this may be a possible reason for more complete combustion. The oxygen molecule present in biodiesel molecular structure may be readily available for oxygen. However, it was noticed that after a certain limit of biodiesel concentration, the thermal efficiency trend is reversed and it starts decreasing. This behaviour of biodiesel fuel needs advanced investigations.

An important observation is that all the blends have a higher thermal efficiency than the baseline data of diesel fuel. A graph between the concentration of ester blend and improvement in peak thermal efficiency for various concentrations of biodiesel blend is plotted in Figure 1.40.

Esterification is a process, which brings about a change in the molecular structure of the vegetable oil molecules, thus bringing down the levels of viscosity and unsaturation of vegetable oils. The viscosity of vegetable oil reduces substantially after esterification. The density and viscosity of the linseed oil methyl ester formed after esterification were found to be very close to petroleum diesel oil. The flash point of LOME was higher than that of diesel oil. The 20 per cent biodiesel blend also demonstrated comparatively higher flash point than petroleum diesel oil and was in range of `safe fuel'. The cetane number for the neat diesel oil obtained by this method is 50 and cetane number of LOME was 52. A 20 per cent blending of LOME with diesel oil improved the cetane number of diesel oil. Lower concentrations of LOME in biodiesel blends can be used as a cetane improver. The calorific value of LOME was found to be slightly lower than petroleum diesel oil. All these tests for characterization of biodiesel oil demonstrate that almost all the important properties of biodiesel are in very close agreement with the diesel oil making it a feasible alternative for the application in compression ignition engines for partial replacement of diesel fuel.

Esterification has been found to be an effective technique to prevent some long-term problems associated with utilization of vegetable oils such as fuel filter plugging, injector coking, formation of carbon deposits in combustion chamber, ring sticking, and contamination of lubricating oils. The carbon deposits on piston top and injector coking substantially reduced in biodiesel-fuelled system. The performance of biodiesel-fuelled engine was marginally better than the diesel-fuelled engine in terms of thermal efficiency, brake specific energy consumption, smoke opacity, wear of vital components, and exhaust emissions for entire range of operations. It was conclusively proved that self-lubricity of LOME in biodiesel played a key role in engine performance. Biodiesel is proved to be a potential candidate for partial substitute of mineral diesel oil.

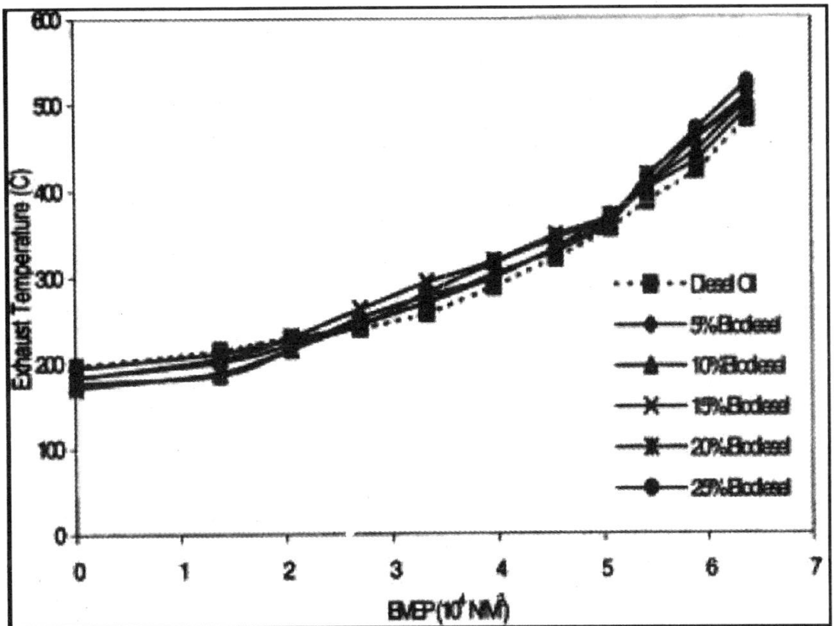

Fig. 1.36. Exhaust temperature Vs *bmep*

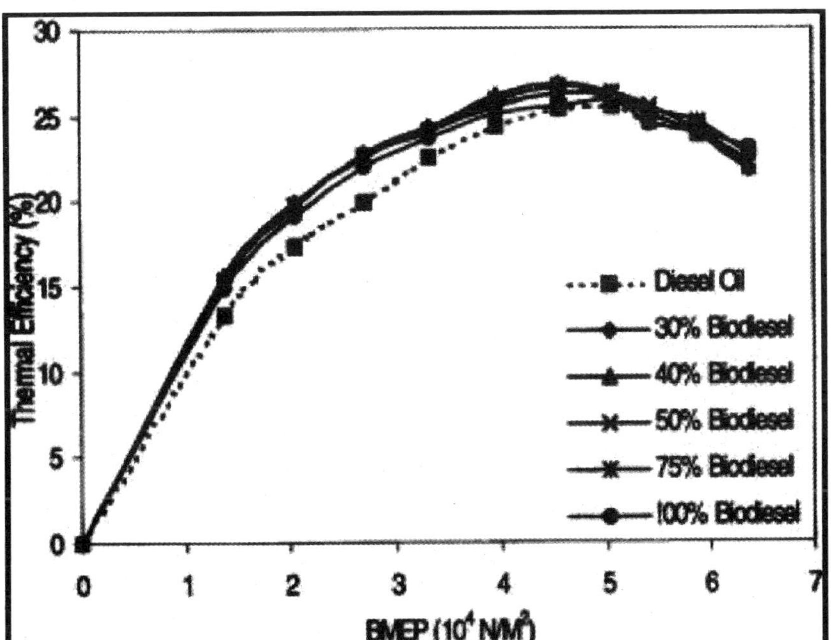

Fig. 1.37. Thermal Efficiency Vs *bmep*

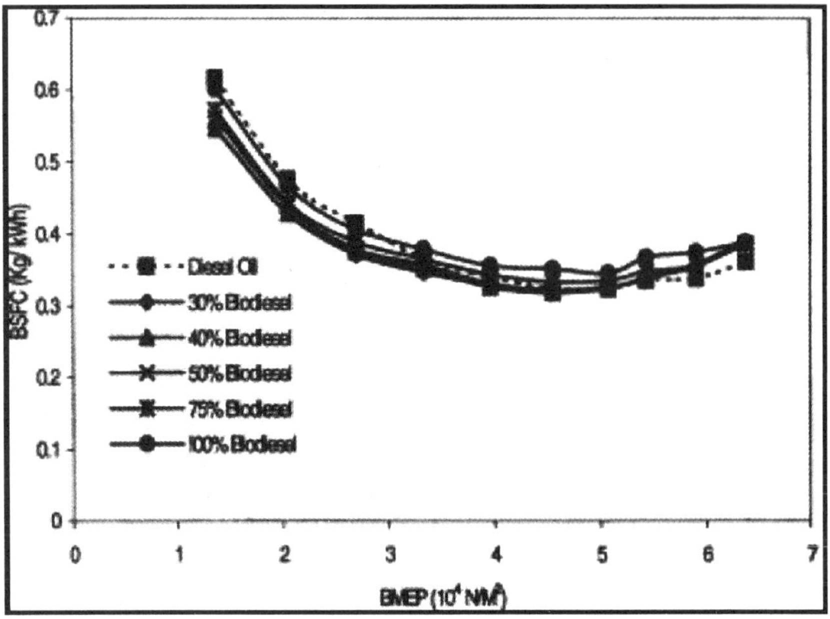

Fig. 1.38. *bsfc* Vs *bmep*

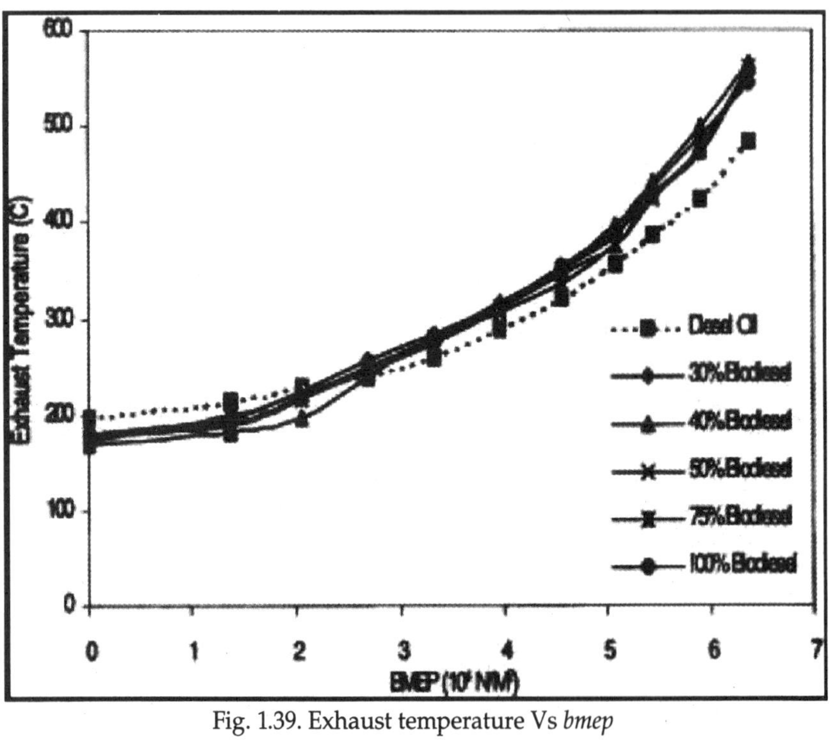

Fig. 1.39. Exhaust temperature Vs *bmep*

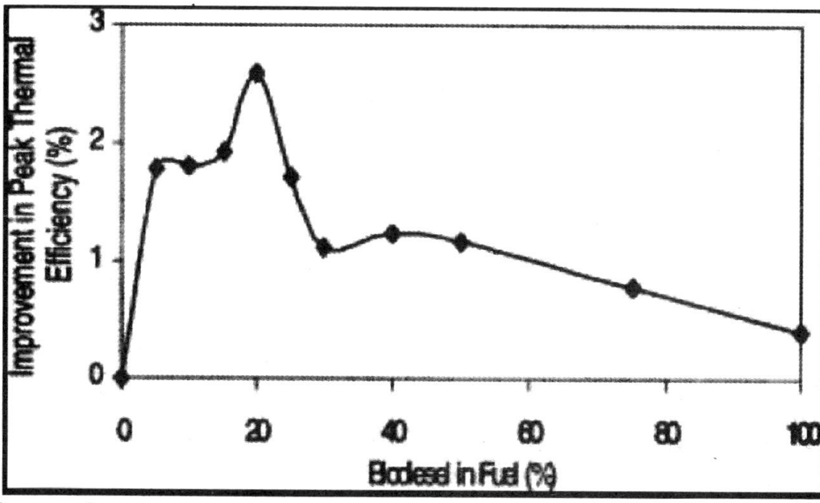

Fig. 1.40. Improvement in peak thermal efficiency with respect to biodiesel in fuel

Studies on Mahua oil methyl ester (MME)

The brake specific fuel consumption (*bsfc*) of each fuel is shown in figure 1.41. The term "brake specific" is used to designate quantities that have been normalized by way of division by the engine's power, thus the *bsfc* is equal to the fuel flow rate divided by the engine's power. Figure 1.41 shows that *bsfc* for MME is higher than diesel. This is due to the fact that ester has lower heating value compared to diesel; so more ester-based fuel is needed to maintain constant power output.

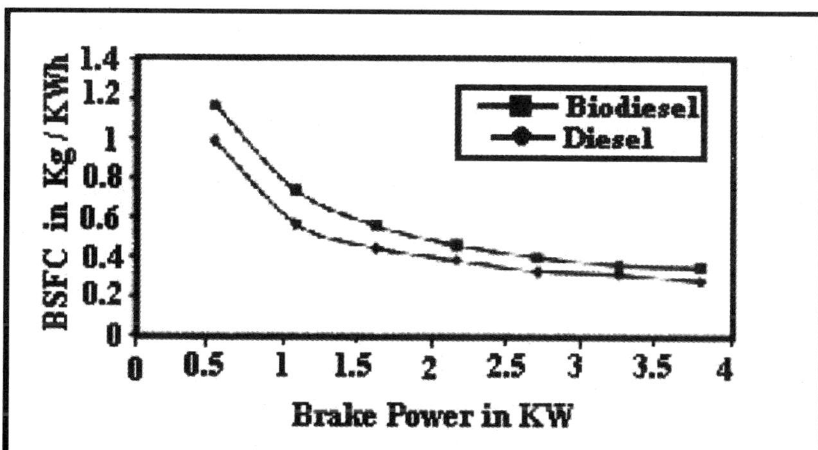

Fig. 1.41. *bsfc* Vs brake power

BRAKE THERMAL EFFICIENCY

Figure 1.42 shows the brake thermal efficiency for ester is lower than diesel fuel. Brake thermal efficiency of an engine depends on a number of factors but when we are discussing about fuel the most meaningful properties are heating value and specific gravity. The combination of heating value and mass flow rate indicates energy input to the engine. This energy input or consumption to the engine in case of MME is more compared to diesel. And it is very clear that the reduction in efficiency is very low compared to diesel fuel, which is due to the oxygen content in the ester, which results in better combustion [66] and [69].

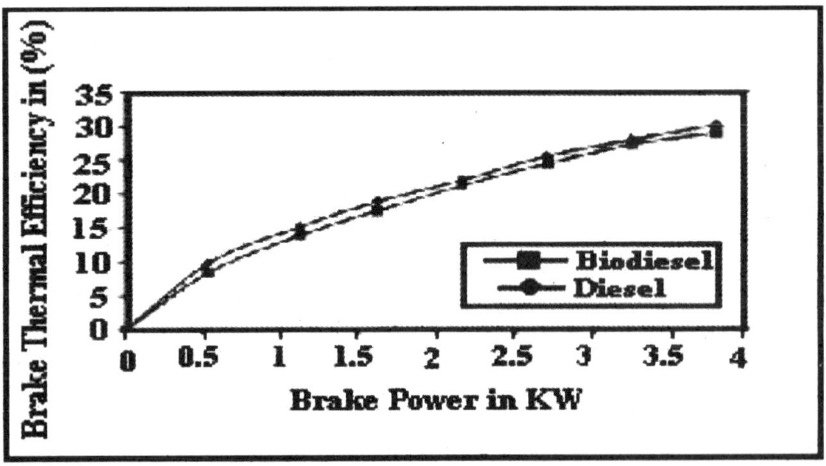

Fig. 1.42. Break thermal efficiency Vs brake power

SPECIFIC ENERGY CONSUMPTION

Figure 1.43 shows that specific energy consumption (SEC) is higher for MME than diesel. The high SEC is due to the lower energy content of the ester. Already there are reports that biodiesel prepared from vegetable oils has high SEC [65], [66], [67] and [68]. The esters that convert the chemical energy to mechanical work are comparable with diesel fuel. This specific energy consumption measures the amount of input energy required to develop one-kilowatt power. This is an important parameter to compare brake thermal efficiency of an engine because it is taking care of both mass flow rate and heating value of the fuel. Here, for MME, both mass flow rate and heating value are different and hence it is simple if comparison is made in terms of specific energy consumption rather than specific fuel consumption.

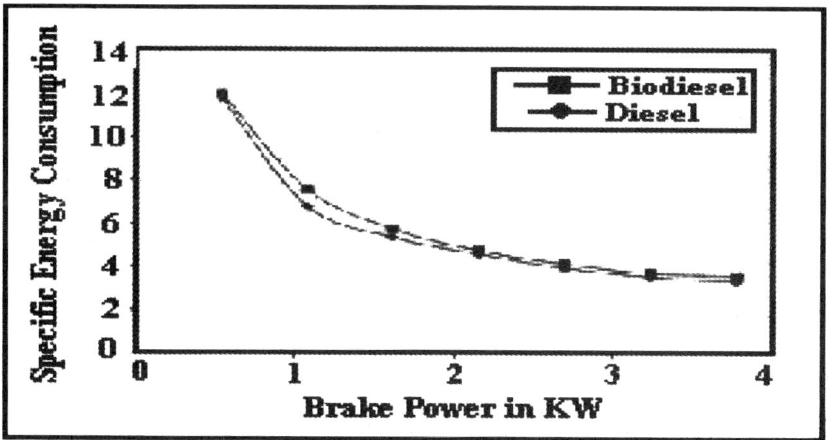

Fig. 1.43. Specific energy consumption Vs brake power

INFERENCE

In this study Mahua oil was transesterified using 6:1 mole ratio of methanol to oil to obtain methyl ester of low viscosity (5.2 CSt) and good conversion percentage (92%). The ester was washed with phosphoric acid to remove traces of alkali and substantially with distilled water. MME possesses lower calorific value around 12 per cent compared to diesel. The specific gravity does not vary much compared to diesel. The kinetic viscosity is slightly higher than that of diesel however within the biodiesel standard limits (5.2 CSt). Cetane number is slightly higher by 10 per cent which are favourable for combustion. Flash and fire points are high which is advantages for fuel transportation. The cloud point is higher than diesel, which creates problem in low temperature regions.

The performance of diesel engine with biodiesel does not vary much. The specific fuel consumption is higher (20%) than that of diesel and thermal efficiency is lower (13%) than that of diesel. Exhaust pollutant emission are reduced compared to diesel. Carbon monoxide, hydrocarbon, smoke number and oxides of nitrogen were reduced 30 per cent, 35 per cent, 11 per cent, 4 per cent, respectively, compared to diesel.

Summary

Characterization of biodiesel depends on the relative merits and demerits and the investigation reports submitted by various authors worldwide in the contemporary time.

Interest depends on how biodiesel is suitable for replacement for petro diesel. It is reasonable to take global parameters into consideration before a sensible recommendation is made. There are no impossible solutions that can impede the replacement without the design changes of the diesel engine to suit the biodiesel. The advantages are overriding as compared to the setbacks biodiesel is suffering from. The setbacks are surmountable with suitable additives or minor affordable design changes. For example, separation of blended constituents in cold weather, increase in NO_x levels by virtue of the constituent oxygen, lube oil dilution in the crankcase, increase in pour point at cold temperatures can be tackled by appropriate remedial measures.

There are inbuilt setbacks that marginally debilitate the biodiesels. For example, lower heating value, higher viscosity, and higher density result in marginal decrease in the performance in some aspects. This can be compensated with the growing trend in the rise of petro diesel price and environmental pollution.

All the methyl esters of the oils tested, especially the non-edible type, can be recommended for replacement because of the reason that the diesel engines can be run without knock and detonation and primarily lesser emissions in the exhaust gases.

In this study, over and above the comprehensive investigations taken up to characterize various biodiesels in the literature, additional parameters like noise and vibration of the engine in different engine running conditions have been taken up. The idea of studying engine vibration and noise in comparison to biodiesel is mooted since the combustion in the cylinder creates the main excitation force and vibration recorded on the cylinder in FFT form is the true reflection of the mode of combustion of particular ester. Time waveforms and phase of vibration on the cylinder are also obtained while the engine is running with the oils to be tested and compared with the diesel baseline signatures.

The idea of studying engine vibration and noise in comparison to biodiesels is mooted to explore the feasibility in an unconventional way to evaluate their suitability in the context of soaring petro-diesel price keeping in view the trend in its rise.

2

Transesterification Procedure

The use of vegetable oils in lieu of diesel fuel in conventional diesel engines requires modification in their properties. Considerable efforts have been made to develop vegetable oil derivatives that would approximate the properties and performance of the hydrocarbon-based diesel fuels. The problem of substituting pure vegetable oils (non-edible) for diesel fuels are mostly associated with their high viscosities. Reduction of viscosity can be effected by any of the processes like transesterification, mineralization, and pyrolysis. Mineralization consumes more time and pyrolysis brings about irregular molecular break down. Hence transesterification of some of the edible and non-edible oils has been taken up in this work to experiment on a laboratory-based engine with the esterified versions and their blends with diesel oil.

Transesterification of Vegetable Oils

Transesterification is the general term used to describe the important class of organic reactions, where an ester is transformed into another through interchange of alkyl groups and is also called as alcoholysis. The transesterification is an equilibrium reaction and the transformation occurs by mixing the reactants. However, the presence of a catalyst accelerates considerably the adjustment of the equilibrium. General Equation for Transesterification Reaction is given below

$$\mathbf{RCOOR' + R''OH \leftrightarrow RCOOR'' + R'OH}$$

The basic constituent of vegetable oils is triglyceride. Vegetable oils comprise of 90-98 per cent triglycerides and small amounts of mono, diglycerides and free fatty acids. In the transesterification of vegetable oils, a triglyceride reacts with an alcohol in the presence of a strong acid or base, producing a mixture of fatty acid alkyl esters and glycerol. The overall process is a sequence of three consecutive and reversible reactions in which di- and monoglycerides are formed as intermediates. The stoichiometric reaction requires one mol of triglyceride and three moles of alcohol. However, an excess of the alcohol is used to increase the yields of alkyl esters and to allow phase separation from the glycerol formed. Several aspects, including the type of catalyst (base or acid), alcohol/ vegetable oil molar ratio, temperature, purity of the reactants (mainly water content in alcohol) and free fatty acid content have an influence on the course of the transesterification [45]. So in this work the reactants of high purity has been used (methyl alcohol with 99.95% purity) in the process. The process employed is the base-catalyzed process, where the transesterification of vegetable oils proceeds faster than the acid-catalyzed reaction [46,47] together with the fact that the alkaline catalysts are less corrosive than acidic compounds.

The mechanism of the base-catalyzed transesterification reaction of vegetable oil, as explained by the Ulf Schuchardt{45}, is shown in the Figure 2.1. The first step (Eq.1) is the reaction of the base with the alcohol, producing an alkoxide and the protonated catalyst. The nucleophilic attack of the alkoxide at the carbonyl group of the triglyceride generates a tetrahedral intermediate (Eq. 2), from which an alkyl ester and the corresponding anion of the diglycerides are formed (Eq.3). The latter deprotonates the catalyst, thus regenerating the active species (Eq.4), which is now able to react with a second molecule of the alcohol, starting another catalytic cycle. Diglycerides and monoglycerides are converted by the same mechanism to a mixture of alkyl esters and glycerol.

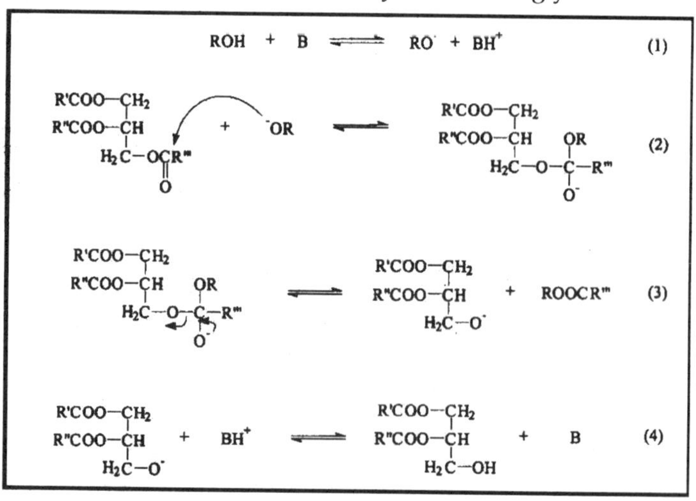

Fig. 2.1. Mechanism of the base-catalyzed transesterification process

Alkaline metal alkoxides (as CH_3ONa for the methanolysis) are the most active catalysts, since they give very high yields in short reaction times even if they are applied at lower molar concentrations. However, they require the absence of water, which makes them inappropriate for typical industrial process. Alkali metal hydroxides (KOH and NaOH) are cheaper than metal alkoxides, but less active. Nevertheless, they are good alternatives, since they can give the same high conversions of vegetable oils just by increasing the catalyst concentration to 1 or 2 folds.

MAKING VEGETABLE OIL METHYL ESTERS

Here one stage base-catalyzed process has been selected for the methyl esterification of non-edible vegetable oils like palm kernel, Jatropha, and Mahua which are forest products available in the region and methanol (99.95% purity) as reactants and NaOH as catalyst. Initially, the process is carried out with 200ml of vegetable oil sample to find out the appropriate quantity of catalyst required for a specific type of oil to get high yields of methyl esters [45]. After successful attempt with 200ml sample, the same proportions of the constituents were used in processing methyl esters of required quantities, for establishing the properties and experimentation on IC engine with a batch size of 1 litre.

PROCESS EMPLOYED FOR MAKING THE METHYL ESTERS

1) Desired amount of fresh vegetable oil is filtered and taken in a reaction vessel.
2) Reaction vessel with the fresh vegetable oil is kept in constant less temperature water bath and allowed to reach 50° C.
3) Twenty per cent methanol per litre of fresh vegetable oil is taken and desired amount of NaOH is weighed and dissolved in it.
4) The above mixture is introduced into the reaction vessel.
5) Now the contents in the reaction vessel are mixed vigorously for 30 minutes.
6) The reaction vessel is removed from heat source and is allowed to settle overnight.
7) The products are checked for the glycerol separation clearly, if not, it is to be treated as failed batch and again a fresh batch with varied NaOH is taken and processed.
8) Upon overnight settlement, separation of dense glycerol layer at the bottom and mixture of methyl ester, catalyst and alcohol as the top layer is observed.
9) Top layer is separated and water washed to remove the excess catalyst and the alcohol, as they are more soluble in water than methyl esters.
10) Washing method employed involves mixing of 50 per cent water and 50 per cent biodiesel taken in a beaker, stirred for few minutes and then allowed to settle for 12 hours.

11) Since biodiesel has a lower specific gravity than water, the water will eventually separate and settle to the bottom along with soluble matters, mainly the remaining alcohol and catalyst, and biodiesel will remain on top of the water.
12) In the first wash the water turns white. The wash water is drained off and the process is repeated until the drained water is totally clear.
13) If the biodiesel is observed slightly cloudy, it is heated to 50°C and is held for 15-20 minutes to clear up the cloudiness.
14) Finally, the biodiesel is allowed to cool and filtered with 10-micron filter paper.

The following flow sheet clearly explains the production of biodiesel from non-edible vegetable oils chosen in our laboratory.

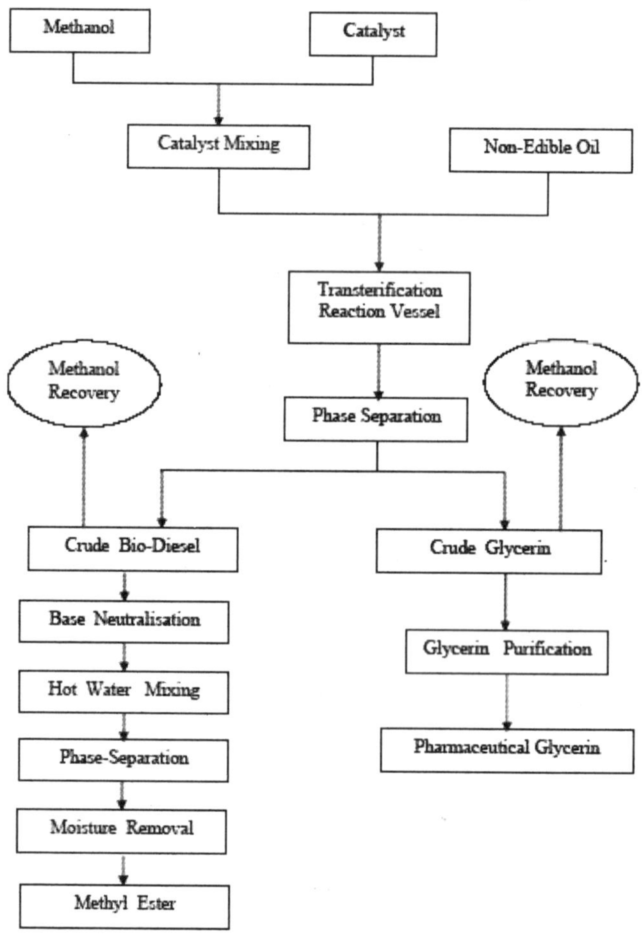

Fig. 2.2. Flow Sheet For Production Of Bio-Diesel From Non-Edible Vegitable Oil

After making the required quantities of methyl esters of each oil' necessary properties of these esters are established as per the IS test methods (IS: 1448). The author has employed the above said process and prepared methyl esters of each oil and the properties of these esters have been established as per the standards recommended by HPCL, Visakhapatnam and Chemical engineering department of A.U. College of Engineering. The results are tabulated in Table 2.1. Later, the esters samples were tested on the laboratory based D.I. Compression Ignition Engine for performance analysis. The Chapter- IV deals with the experimental setup and the experimental procedure employed.

S. No	Name of the oil sample	Diesel	Jatropha	Jatropha Methyl Ester	Palm kernel	Palm Kernel Methyl Ester	Mahua	Mahua Methyl Ester
1.	Density at 30^0C Kg/m^3	830	918.6	880	916	875	875	860
2.	Gross Calorific Value, KJ/Kg	43000	39770	38450	39250	41500	30248	38650
3.	Viscosity at 30^cC	4.59	49.93	5.65	63.6	4.5	16.9	5.2
4.	Cetane Number	45-55	40-45	50	40-45	50	40-45	48
5.	Rams bottom Carbon Residue, Wt%	0.1	0.64	0.5	0.45	0.38	0.46	0.15
6.	Flash Point, ^0C	50	240	170	200	172	100	171
7.	Pour Point, ^0C	Winter 3 Max, Summer 15 Max	20	<-3	27	<-3	27	<-3
8.	Acid Number, mg KOH/gm	0.2 Max	10	<0.2	1.2	<0.2	20	<0.2

Table 2.1. Properties of diesel, non-edible vegetable oils and their methylesters

The following Chapter (Chapter III) deals with the mathematical modelling for the calculation of the heat release rate calculations based on the first law of thermodynamics.

3

Heat Release Rate Calculations

This chapter deals with the mathematical modeling for the calculation of the heat release rate calculation based on the first law of thermodyamics

Heat release analysis of the engine reflects the combustion process occurring in the engine. It is carried out within the framework of first law of thermodynamics. The combustion, heat transfers and mass laws affect the pressure crank angle data of an engine. Thermal efficiency and peak cylinder pressure are influenced very much by the combustion process. Heat added before TDC increases heat losses, frictional losses and loss of peak cylinder pressure. Advancing the combustion not only leads to increase in effective expansion ratio but also increases the previously mentioned losses. The crank angle at which 50 per cent of total heat is released has been identified as an effective parameter for online engine tuning [48]. Heat release equations can be checked and modified suitably by comparing them with the heat release rate curves obtained from engine pressure vs. crank angle data. This enables the heat release equations to be used in cycle simulations, which can result in considerable savings in cost and time during engine development.

The different methods for computation of heat release rate from cylinder pressure data vary in the degree of accuracy with which the contents of the cylinder are considered. Some methods are simple and easy to use and some others are complicated and involve extensive computation to achieve accuracy. Krieger et al. [49] strives for extreme accuracy in representing the thermodynamic properties of the working medium. This leads to complicated computations. The model needs wall temperature and heat transfer coefficient estimates and has to be used along with a detailed cycle simulation. Gatowski et al. [50] developed a one zone model in which the specific ratio of the working medium is represented by a linear function of temperature. They have included crevice effects. Wall temperatures and heat transfer coefficients are estimated. Hayes et al. [51] estimated the heat release rates of a diesel engine by assuming air as working medium. The model needs initial estimate of mass of the cylinder contents through a cycle simulation and heat transfer estimates. Karim et al. [54] estimated the heat release rates in duel fuel engines using a model that considers dissociation of the products of combustion and needs heat transfer estimates and discussed a method, which assumes that the cylinder charge has a ratio of constant pressure to constant volume. Specific heat ratio is equal to the average measured polytropic index during compression and expansion as the case may be. Hence the heat transfer is not considered separately. Varaprasad Rao [52] implemented a scheme, which uses the average polytropic index of compression and expansion to calculate the cumulative heat release to exhaust valve opening, excluding the period between occurrence of the spark and the end of combustion. Heat release analysis, which can be easily computed with a simple programme, is presented. In his study, instantaneous compression and expansion indices are used to calculate heat release rate based on the equations developed by Ramesh et al. [53]. This method of calculating the exponents of compression and expansion directly eliminates the work of plotting log (p) vs. log (v) and estimating the mean exponents graphically. During the combustion period, the polytropic exponent is varied linearly from the average value during compression to that during expansion.

Heat Release Equation

Considering the cylinder charge to be homogeneous, from the first law of thermodynamics,

$$\Delta Q_c = \Delta W + \Delta Q_h + \Delta U \tag{1}$$

where Q_c = Chemical Energy Released in KJ/Kg,

W = Work Output in KJ/Kg,

Q_h = Heat Transfer from the cylinder in KJ/Kg,

U = Internal Energy of the contents of the cylinder KJ/Kg,

Assuming the cylinder contents to be an ideal gas,

$$\Delta Q_c = P.\Delta V + mC_v.\Delta T + \Delta Q_h \tag{2}$$

where

P = Cylinder Pressure in bar,

V = Cylinder Volume in m^3/kg,

m = mass of the cylinder charge in kg,

C_v = Specific heat at constant volume of the cylinder contents in KJ/Kg^0 K,

T = Charge Temperature in K

Eq. (2) can also be rewritten as

$$\Delta Q_c = P\Delta V\left(\frac{\gamma}{\gamma-1}\right)+\left(\frac{1}{\gamma-1}\right)V\Delta P+\Delta Q_h \qquad (3)$$

Where γ = ratio of specific heats.

The rate of change of chemical energy released with respect to crank angle may be expressed as

$$\frac{\Delta Q_c}{\Delta\theta} = \left(\frac{n}{n-1}\right)P\frac{\Delta V}{\Delta\theta} + \left(\frac{1}{n-1}\right)V\frac{\Delta P}{\Delta\theta} \qquad (4)$$

Where n is the polytrophic exponent and θ is the crank angle. This equation takes care of heat transfer effects also. During polytropic compression expansion,

We can write (5) $P_1 V_1^{\ n} = P_2 V_2^{\ n}$

This leads to the expression $n = \dfrac{\ln\left[\dfrac{P_1}{P_2}\right]}{\ln\left[\dfrac{V_2}{V_1}\right]}$ (6)

Equations (4) & (5) are used to calculate the heat release rate. Applying equation (6) between successive data points, the instantaneous polytrophic exponent is obtained. The heat release rate is then integrated to give the cumulative heat release pattern.

Pressure readings are obtained from a pressure pick-up flush mounted on the cylinder head wall and crank angle encoder recorded crank angle readings. These two outputs were integrated by an engine recorder as shown in Fig.4.2, and presented as p-θ diagram with suitable software. The practical data acquired for one particular cycle of operation is presented in Appendix 'C'. On substitution of the experimental data in the governing equations presented above, the heat release rate has been computed. A derived heat release rate curve from the above method for the engine running at full load, and at rated rpm of 1500 is presented in Fig.3.2. Pressure and crank angle values at this load and speed are used as inputs for the heat release compilation through Finite difference method. The experimental data recorded is comprehensively presented in the graphical plots, (Figures 3.1-3.3) in which information is used for evaluating the heat release rate under different operating conditions. The similar kinds graphical plots for various percentages methyl esters at various load conditions are presented in Appendix C.

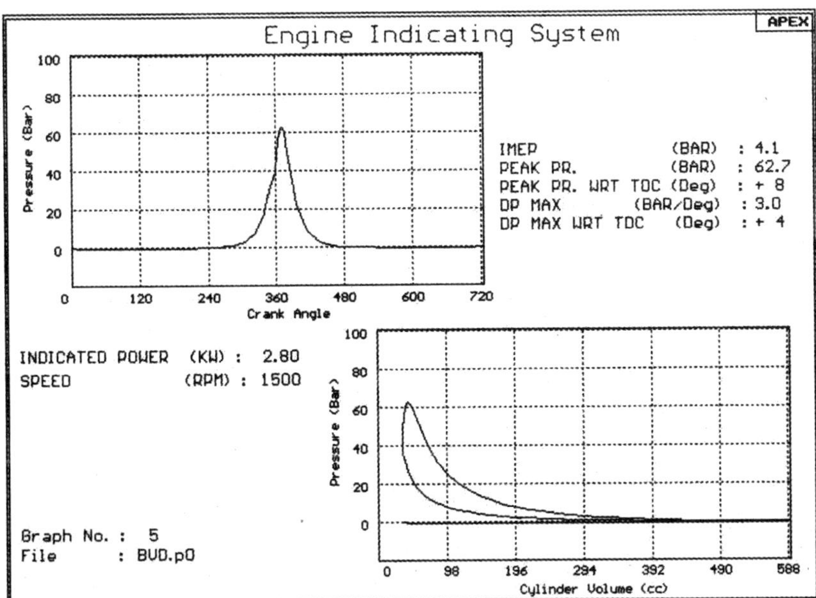

Fig . 3.1. P-θ & PV diagram for the engine running on pure diesel at full load

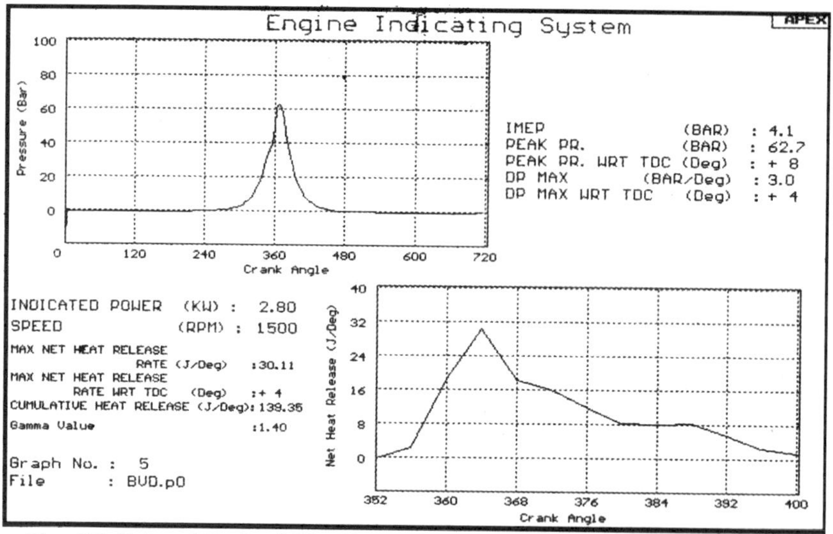

Fig . 3.2. P-θ and Derived Net Heat Release curves for pure diesel full load

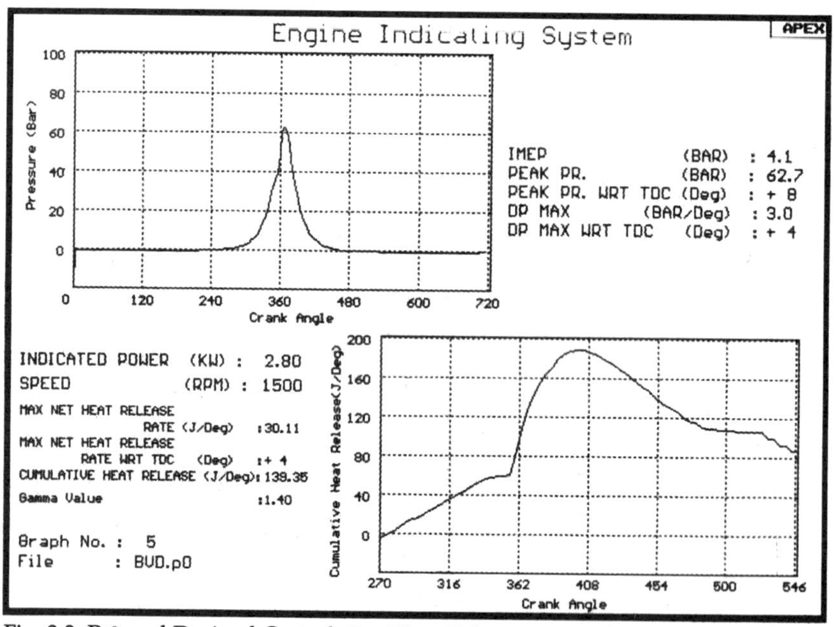

Fig. 3.3. P-θ and Derived Cumulative Heat Release curves for pure diesel at full load

4

Experimental Set Up

The experimental setup of DI diesel engine is arranged in the engines laboratory. Experimentation is carried out at various engine loads to study the cylinder pressure and heat release rate with respect to the crank-angle. Engine performance data is acquired for the esters as well as pure diesel oil independently to study the above-mentioned parameters. Engine cylinder vibration is monitored at each load and for each ester simultaneously to compare the cylinder excitation frequencies with the base line frequencies using diesel oil. Since the very combustion in the cylinder is the basic exciter, the vibration study of the engine cylinder through the derived FFT and overall levels is the representative of combustion propensity.

Experimental set up

The schematic diagram (Fig. 4.1) represents the instrumentation set up for the experiment. The Piezo electric transducer is fixed (flush in type) to the cylinder body to record the pressure variations in the combustion chamber. Crank angle is measured through the crank angle encoder and a plot is drawn between the pressure and crank angle, which is the basic signature to compare the performance of the engine run with various oils. Accelerometer is mounted on the cylinder head to record the engine vibrations using DC-11 data logger, which directly gives the spectral data in the form of FFT, the overall vibration levels and the phase of vibration. Eddy current dynamometer is coupled to the engine for loading purposes. The engine data logger is commissioned to collect the data from the pressure transducer and crank angle encoder and to transmit the same to the computer to display the data in graphic form. These two signals collected are integrated and a pressure-crank-angle plot is compiled with suitable software C7112. The total arrangement of the instrumentation with respect to engine position is shown in the diagram 4.1 and the engine data logger circuit is depicted in the schematic diagram shown in Fig. 4.2.

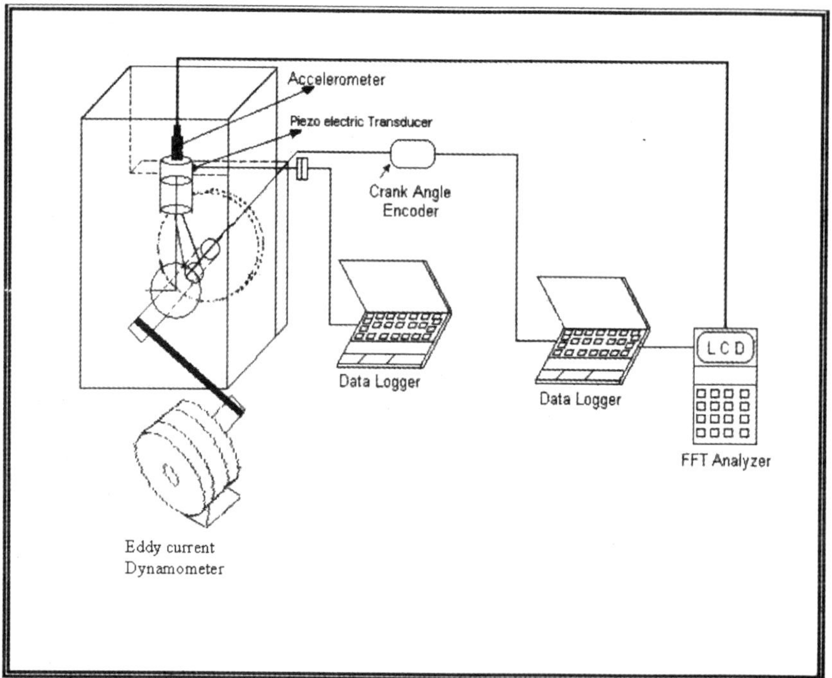

Fig. 4.1. P-θ and Derived Cumulative Heat Release curves for pure diesel at full load

Fig.4.2. depicts the schematic diagram of the connections from the piezo electric transducer and the crank angle encoder to the data logger. The data from the above said transducers is translated into graphic form to construct p-θ diagram.

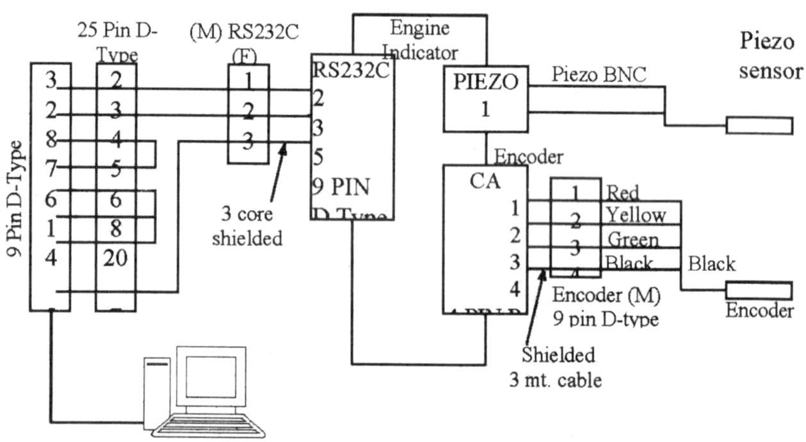

Fig. 4.2 Schematic diagram of Data Integration Circuit taking data from the encoder and the pressure transducer

DIRECT INJECTION (D.I.) DIESEL ENGINE

The DI diesel engine (make Kirloskar Company, Pune) was used for conducting the experimentation. The details of the engine are as mentioned below:

Rated Horsepower: 5 hp

Rated Speed: 1500rpm

No of Strokes: 4

Mode of Injection: DI

No of Cylinders: 1

ENGINE LOADING SYSTEM

In order to load the engine, Eddy current dynamometer is used and is essentially an electric brake, which resists rotation. This rotational resistance is what forces the engine to work during the test. The engine and dynamometer are connected by a belt drive arrangement as shown in Figures 4.1 and 4.2.

Fig. 4.3 Diesel engine test rig

Fig. 4.4. Engine loading device Eddy current dynamometer with Spring balance

Fig. 4.5. Vibration readings on the engine Eddy current dynamometer's operating panel can be seen in this Figure

Fig. 4.6. Noise measurements on the engine with Larson-Davis labs model 710 Dosimeter

Fig. 4.7. Connection from crank angle encoder fixed to the engine

Fig. 4.8. The engine data logger in interface with the computer and the data being logged by the developed software C7112

Eddy Current Dynamometer Details

Speed: 1000-2000 rpm

Peak torque at: 1000 – 2000 rpm

Accuracy ratings +/- 0.3 % to 0.5 %full scale

Type of Loading: Eddy Current Dynamometer with compatible 24 KG representing the Max Load on the engine with a suitably designed spring balance.

PIEZO ELECTRIC TRANSDUCER (S111A22, SN9982)

A Piezo-electric transducer [Fig. 4.9] is fixed on the engine cylinder head for obtaining the combustion chamber pressure data at regular intervals and transmits the same to the computer on-line. Due to the Piezo-electric effect, these mechanical properties manifest themselves as electrical equivalent properties.

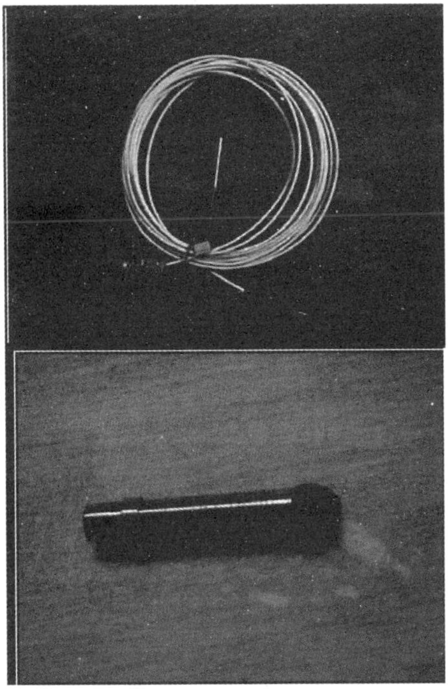

Fig. 4.9. Piezo Electric Transducer

VIBRATION ANALYZER EQUIPMENT

The DC-11 FFT (Fig.4.10) analyzer is a fully digital spectrum analyzer and data collector specifically designed for machine condition monitoring, advanced bearing fault detection, and measurement diagnostics. This instrument measures and displays the following along with rotary machine component balancing as an additional facility. The vibration accelerometer and the optical stroboscope are depicted in the Figures 4.11 and 4.12 respectively.

a) Time wave form (oscilloscope)

b) FFT auto spectra

c) Envelop spectra selected by multiple band pass filters

d) Rotation speed

e) Amplitude and phase on rotation speed and its harmonics

FFT Analyzer Details :(Fig.4.10)

Frequency Range: 1-2000 Hz

Input Signal Range: 100mv

Gain: Auto, 0 -54 dB in 6 dB steps

Input parameters:

Frequency Span: 1-2000 Hz in 1Hz resolution

Frequency Resolution: 1600 lines

Signal to Noise Ratio: Greater than 70 dB

Linear Averages: 1 – 256

Weighting Function: Hanning

Pass filters: None

Amplitude Measurement Units: Acceleration, velocity and displacement

Scale: Peak values

Data Storage Capacity: 400 line spectra 700, 800 line spectra 400, 1600 line spectra200.

Fig. 4.10. DC-11 Vibration data logger

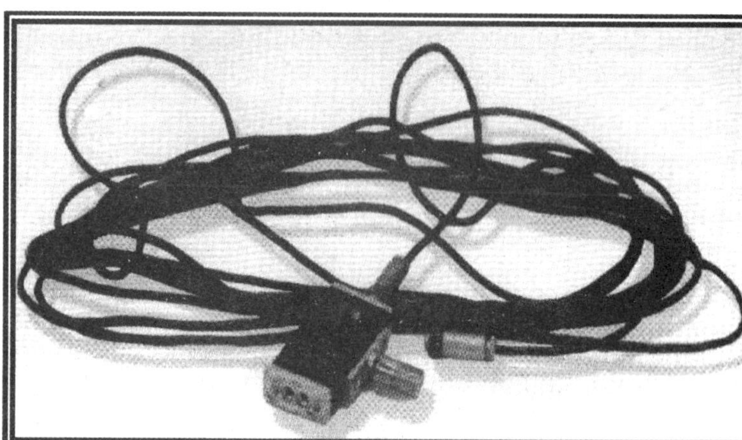

Fig. 4.11. Optical Stroboscope

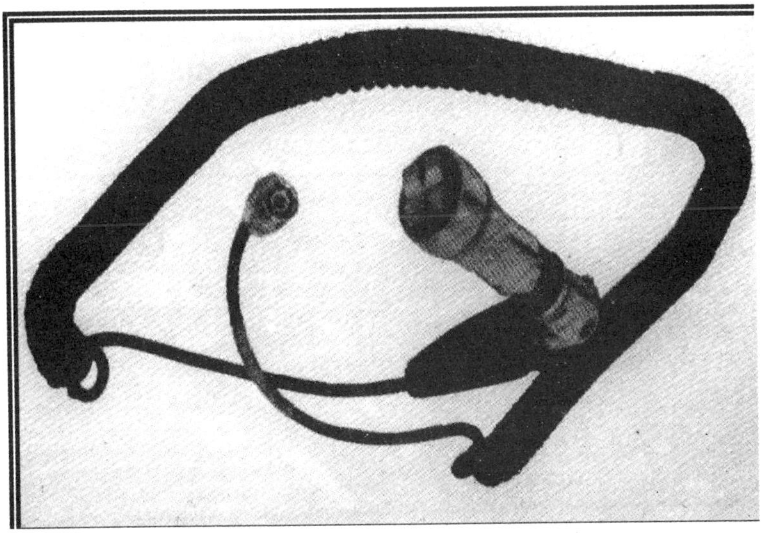

Fig. 4.12. Piezo Electric Accelerometer

DATA LOGGING EQUIPMENT FROM THE ENGINE CYLINDER AND SOFTWARE:

C7112 is a laboratory engine data acquisition system software and is the enterprise-wide software environment designed to support all aspects of telemetry data acquisition, processing, archiving, and display. It gathers the information about a system or process from the data recorder. It is a core tool to the understanding and control of such systems or processes. A fully shielded wiring loom, complete with sensors and probes to accept data and the parameters information such as crank angle and pressure, gathered by sensors, converted to digital format and forwarded onto a computer online for analyzing purposes as shown in Fig 4.2. Combining real and non-real time system services with secure network architecture, captured pressure-crank angle data as input is displayed in real time into the LCD display of laptop in this experimentation. (Fig. 4.8)

EXPERIMENTATION PROCEDURE

The experimentation is conducted on the engine operated at normal room temperatures of 28 to 33°C and using various neat esters of non-edible oils and diesel oil at five discrete part load conditions. The data collection is done independently for the above said oils. The engine is allowed to run at 1500rpm continuously for one hour in order to achieve the thermal equilibrium under operating conditions. After this period, cylinder pressure and crank angle are monitored for every load on the engine. The engine vibration is measured at strategic points on the cylinder and on the mounting of the engine to evaluate the relative performance in the context of vibration emanated by the engine. P-θ and heat release rate graphs have been derived for the above said esters with the software based on the first law of thermodynamics.

SUMMARY

The fuel consumption for all the esters as well as the diesel is measured at all defined loads. This is an attempt to evaluate engine performance for comparison, which is taken up in chapter V.

The indicated power is obtained from the measurements and compilation of the combustion pressures with respect to the crank angle in the cylinder at the defined loads. An algorithm is developed to calculate the mean effective pressures at every load.

The quantities like net heat release rate and cumulative heat release rate are derived from the pressure-crank angle signatures by the software developed.

The engine vibration is monitored by assessing the vibration on the cylinder in three different directions as well as on the engine mounting. The time waveforms and phase of vibration are also measured on the cylinder while the engine is running. The results are elaborately discussed in the chapter five.

5

Results Discussion and Conclusions

Data pertaining to the engine performance, noise and vibration is collected from the engine by using diesel oil as base line oil and then using the methyl esters of various non-edible oils viz. Jatropha, Mahua and Palm Kernel. These three oils are the non-edible oils freely available in the region either in the local forests or in the semiarid lands. The author feels it necessary to validate the performance data with the vibration and noise data emanating from the engine by the implementation of the esters in neat form without changing the engine design. Any change in the vibration in excess which the engine develops with the conventional diesel needs an attempt to verify the suitability of the biodiesel or modify the design of the engine to reduce the vibration. Since the combustion and mass effects are changing with the change of oil, which obviously alters the structural excitation, it is thought to investigate the excesses in the vibration and noise emitted by the engine. Combustion in the cylinder and the flame propagation takes place with certain finite number of frequencies (albeit larger frequency range) in the heterogeneous combustion. Since these burning frequencies form the excitation frequencies of the cylinder, vibration measured on the cylinder in FFT mode reflects the combustion propensity of the type of oil used. In that way, the performance of biodiesel oils is correlated with that of the petro-diesel to understand the replacement aspect conclusively. This method is mooted especially in the event of the conventional performance quantities differ by dismissible margin and commonality in many aspects is being established.

The data collected makes three parts.

- Combustion pressure data and the fuel consumed by the engine at various loads for all the oils under consideration.
- Vibration of the engine on the cylinder and on the foundation, especially by measuring the acceleration amplitude.
- Sound pressure levels at one-meter distance from the engine at various loads.

The pressure-crank angle data is collected for all the oils under consideration (non-edible oil methyl esters and diesel oil as baseline oil). The data is collected at various loads (five load points viz. no load, quarter load, half load, three quarters load and full load). Pressure-crank angle data is the data collected by the engine recorder, which converts the pressure data into the graphic form by using C7112 engine software. This data forms a baseline acquisition to derive other parameters like pressure–volume, log **p** -log **v**, net heat release rate and cumulative heat release rate based on the First Law of thermodynamics, which is discussed in the chapter – III.

As mentioned above, the vibration data is collected and reported in the form of spectrum averages and overall levels and is presented in the graphic forms (FFT and time wave forms by using Vast-an and On-Time software) for comparison and evaluation. Even though the vibration study cannot be directly connected to the engine performance, it can be a suitable tool to assess the occurrence of knock and detonation, if any, during combustion. The amplitude and frequency change in the burning rate brings enough variation in the vibration propensity recorded on the cylinder.

The combustion pressure data is collected at all the loads for the oils under consideration and compared graphically with the diesel oil (Figs. 5.1-5.5). The individual P-θ plots are also drawn for the oils (Figs. 5.8-5.11) and it can be assessed that the diesel, JME and PKME exhibited steep rise in combustion pressure with marginal variation in peak pressures. It can be observed from the individual pressure plots that the pressure rise trends for the diesel and JME are similar with higher peak pressure in the case of diesel oil. Similarly, the pressure plot trends for the MME and PKME are almost same as can be observed in Figure 5.5a.

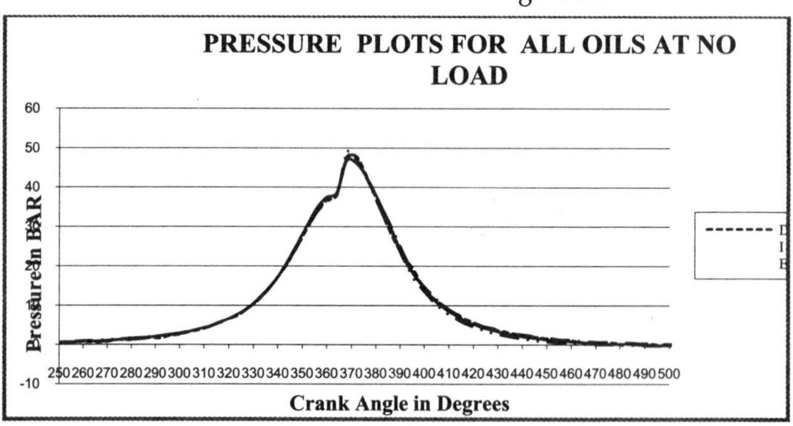

Fig. 5.1. Comparison of combustion pressures for all oils at No-Load

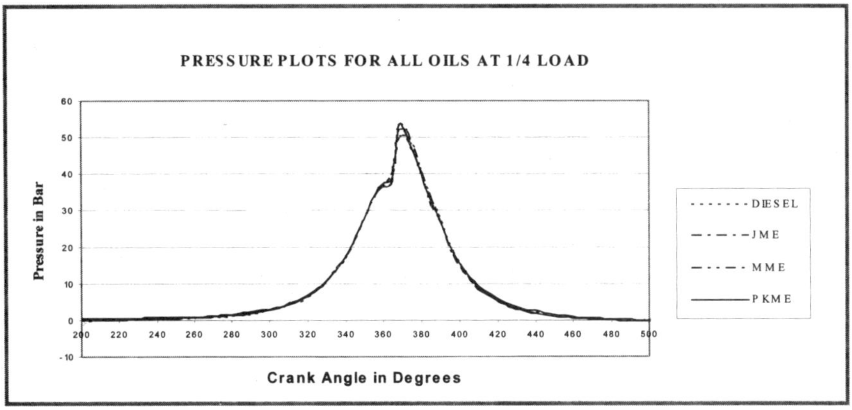

Fig. 5.2. Comparison of combustion pressures for all oils at ¼ Load

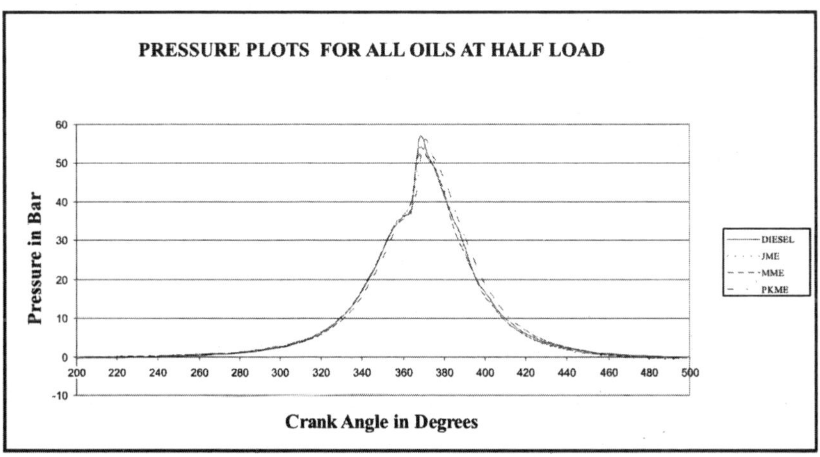

Fig. 5.3.Comparison of combustion pressures for all oils at Half-Load

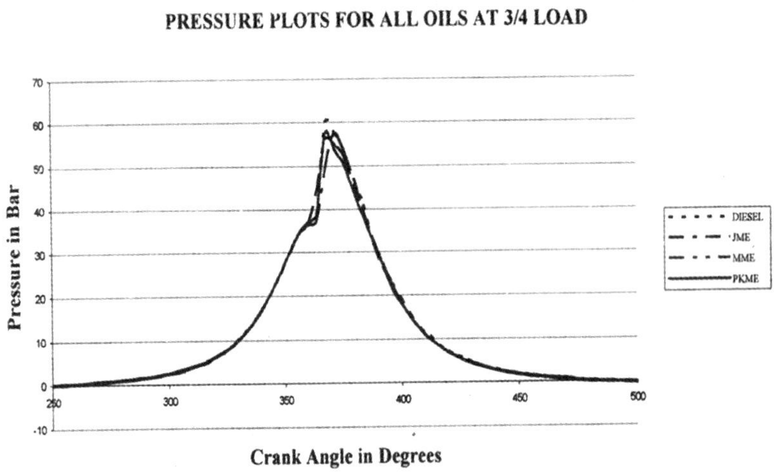

Fig. 5.4. Comparison of combustion pressures at 3/4th – Load

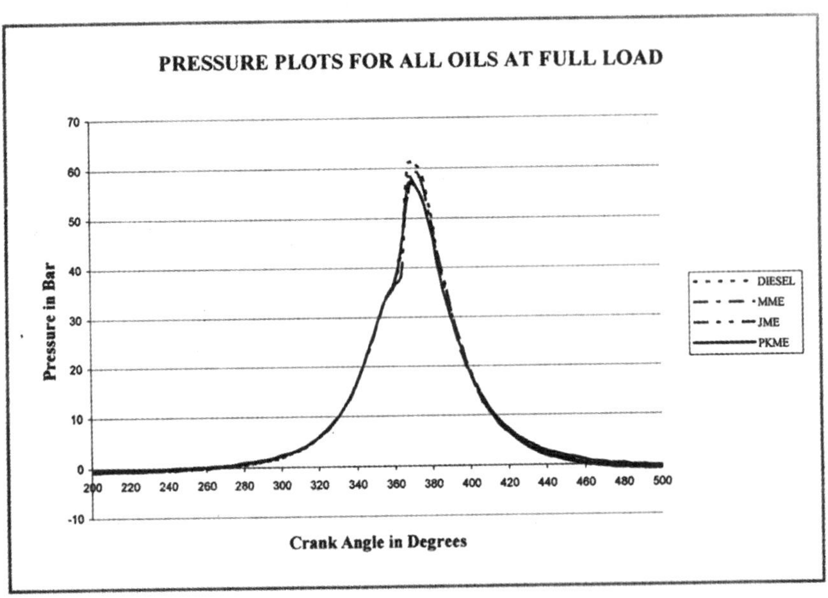

Fig. 5.5. Comparison of combustion pressures for all oils at Full- Load

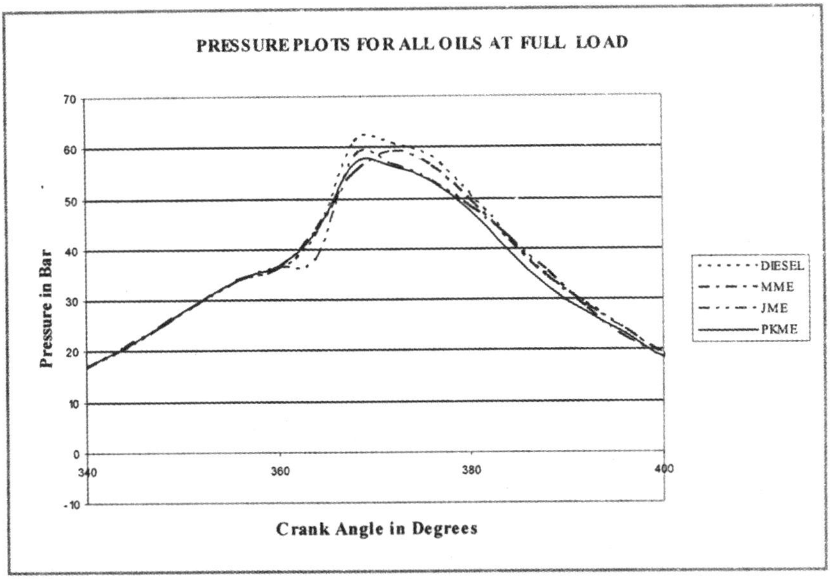

Fig. 5.5a. Comparison of combustion pressures for all oils at Full- Load

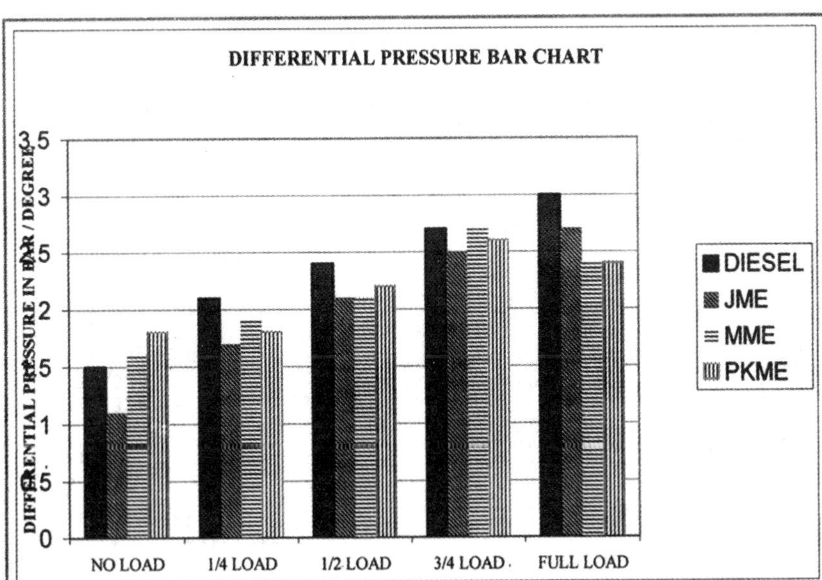

Fig. 5.6. Comparison of differential pressures for all oils at all Loads

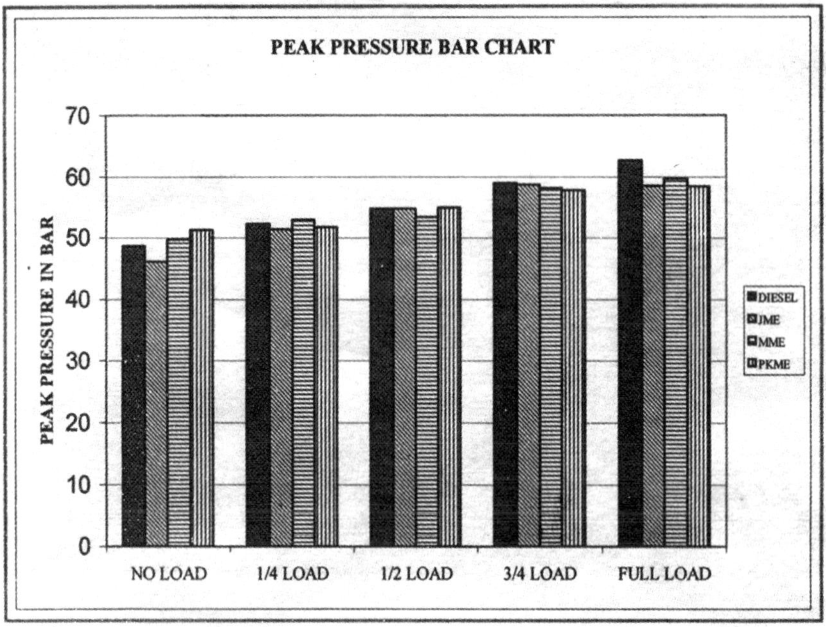

Fig. 5.7. Comparison of peak pressure for all oils at all Loads

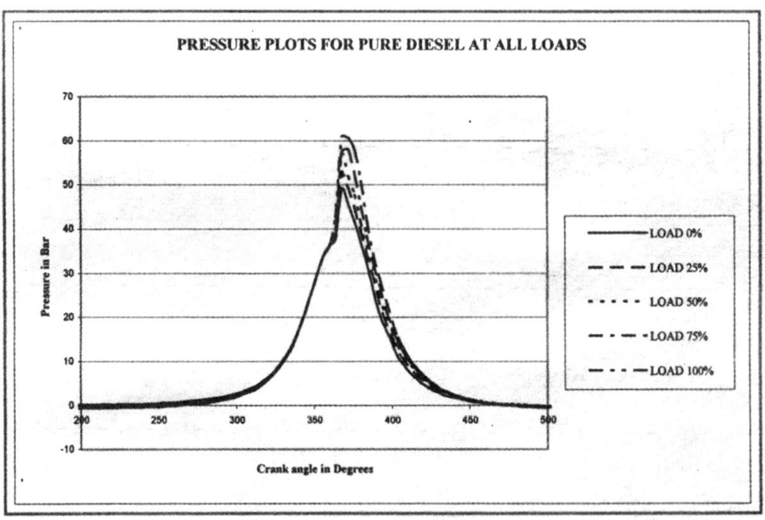

Fig. 5.8. Comparison of combustion pressures at all loads with pure diesel run

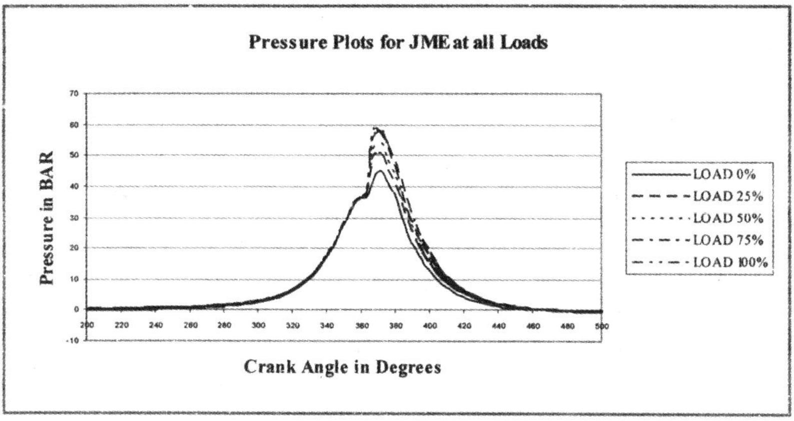

Fig. 5.9. Comparison of combustion pressures at all loads for JME run

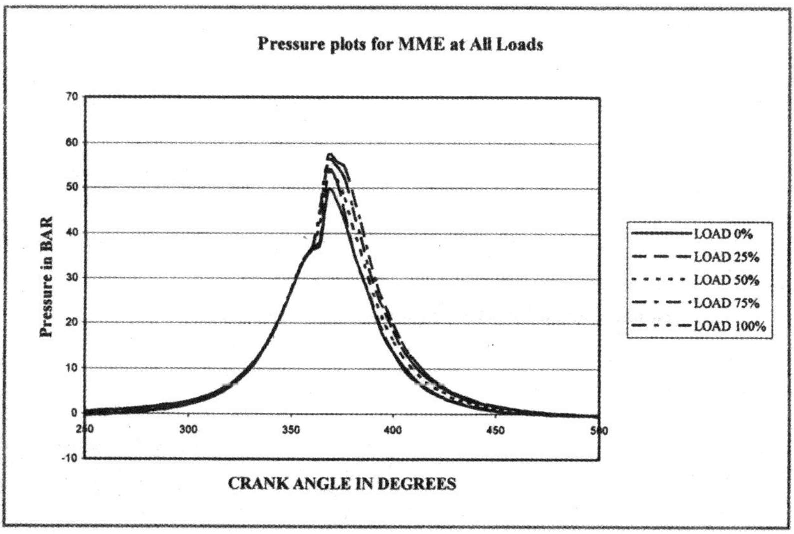

Fig. 5.10. Comparison of combustion pressures at all loads for MME run

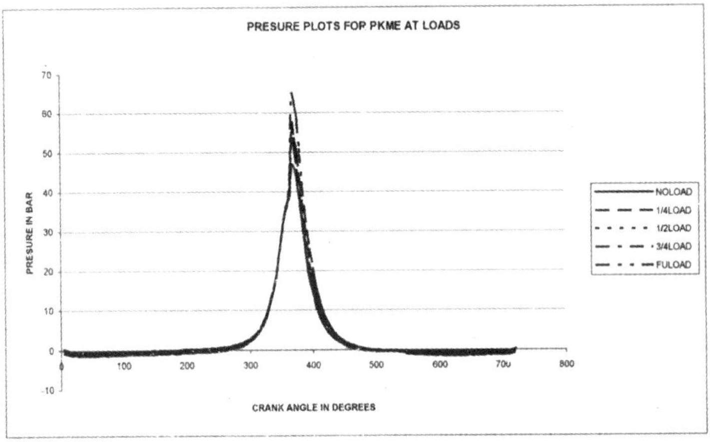

Fig. 5.11. Comparison of combustion pressures at all loads for PKME run

Performance Studies

The performance curves of the engine are drawn as shown in Figs. below (Figures 5.12-5.22). MME stands first in fuel consumption at all loads as per the experimental investigation. The consumption is more by 15-30 per cent, followed by PKME and then JME, and last the diesel oil, which is consumed minimum at all loads. The consumption of methyl esters of the three non-edible oils converges to the same consumption at the full load (Fig. 5.12). Brake specific fuel consumption follows suit of the absolute fuel consumption. Mechanical efficiency of the engine run with the methyl esters of non-edible oils is greater than the diesel by 10-15 per cent with an exception of PKME developing almost the same efficiency at higher loads comparable with the diesel. Indicated thermal efficiency and brake thermal efficiency at all loads are more for the diesel whereas with the methyl esters the engine is generating lesser thermal efficiency by 5-10 per cent, (Figs. 5.15 &5.16). The cumulative heat release rate is coinciding at higher loads for all the oils' and at full load PKME leads marginally in this value. The heat release rate curves are almost same for the diesel oil, PKME and JME with the trends in premixed combustion and the diffused combustion zones are similar, whereas MME's premixed zone is split into two rates and with the diffused combustion zone similar to the pure diesel (Fig. 5.21). The specific gravities and the calorific values of the esters are varying by some degree when compared to the diesel oil and this aspect stands as the main reason for the above said variation in performance. Figures 5.6 and 5.7 represent the maximum differential pressures and peak pressures generated at various loads for the three different esters and the petro-diesel. Petro-diesel is developing highest peak pressures and the differential pressures indicating its lesser ignition delay when compared to the esters tested. The esters tested are sluggish in generating the above aspect with more ignition delay. The sound emanated (Figs. 5.23, 5.5a) is the standing evidence for the above discussion on maximum differential and peak pressures. The Figures from 5.19 to 5.22 represent the net heat release rate curves pertaining to the three esters and the petrodiesel. All the esters behaved similar to diesel except the MME, since the premixed combustion is dull for the MME, which has taken place in two rates, and the diffused combustion is almost flatter comparatively. This is the reason MME is generating consistent overall vibration levels (Fig. 5.24) at all the loads and smooth pressure distribution curve in Figure 5.5a. Ester of Mahua maintained sustained combustion without violent trend in premixed combustion stage despite the reason it generates lower pressures and power comparatively. In addition to this finding, MME's performance has suffered at higher loads as can be observed from the Figures 5.12 to 5.18. The diffused combustion phase is better than the premixed combustion, which is responsible for the higher frequency generation as can be observed from the Figure 5.35. The diffused combustion phase is synonymous with the high frequency generation in the cylinder vibration which conspicuously appeared in the vibration FFT recording on the cylinder perpendicular to the crank in radial direction.

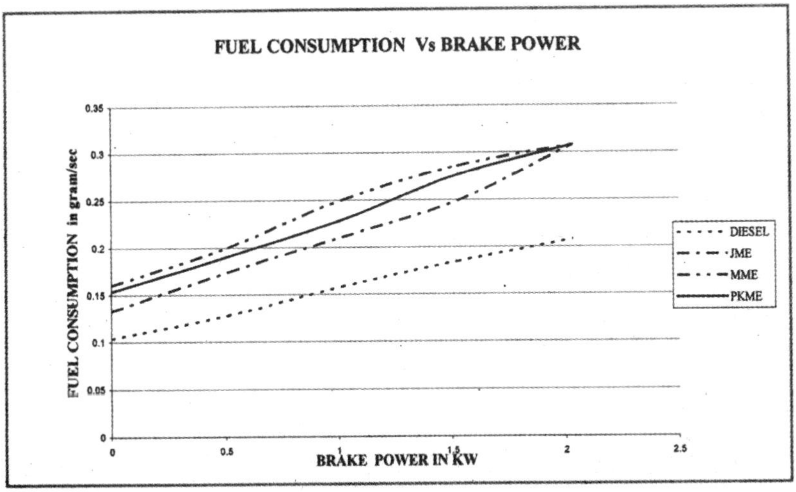

Fig. 5.12. Comparison of brake power in KW Vs fuel consumption in grams per second

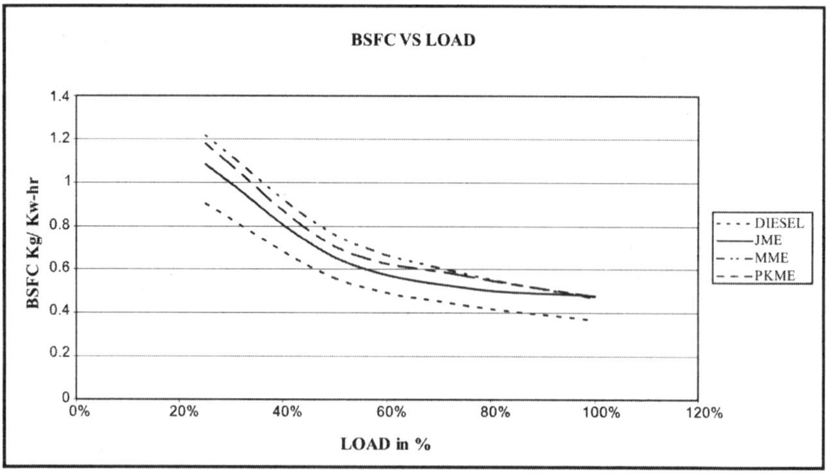

Fig . 5.13. Comparison of load per cent Vs brake specific fuel consumption in kg/kw-hr

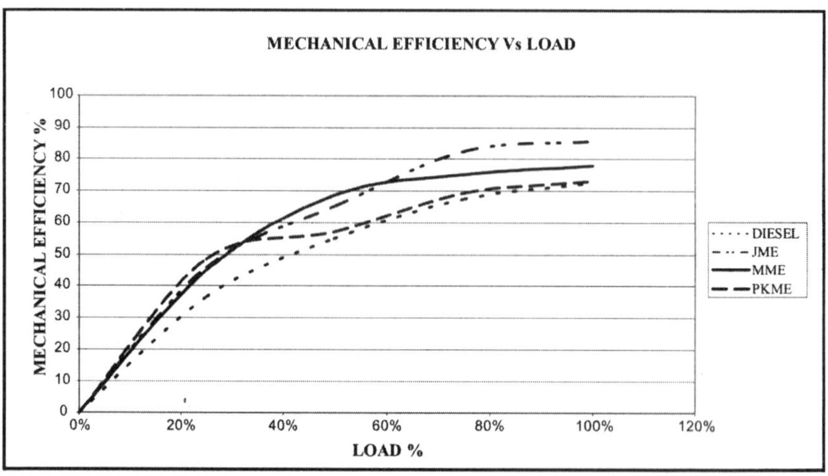

Fig. 5.14. Comparison of mechanical efficiency Vs load per cent

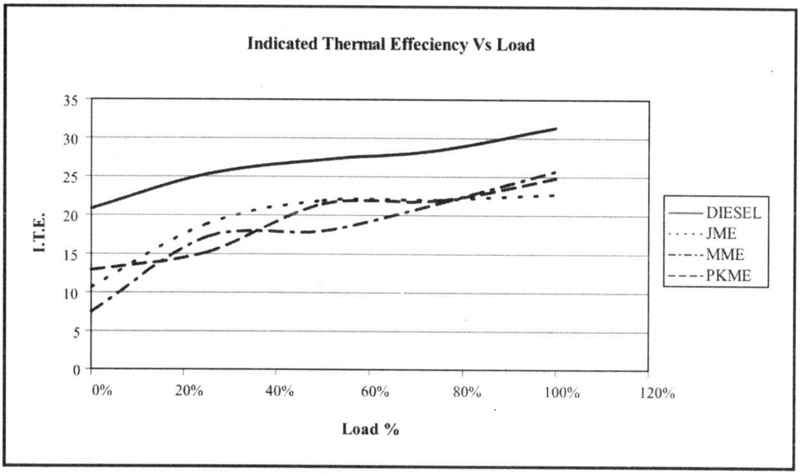

Fig. 5.15. Comparison of indicated thermal efficiency Vs load per cent

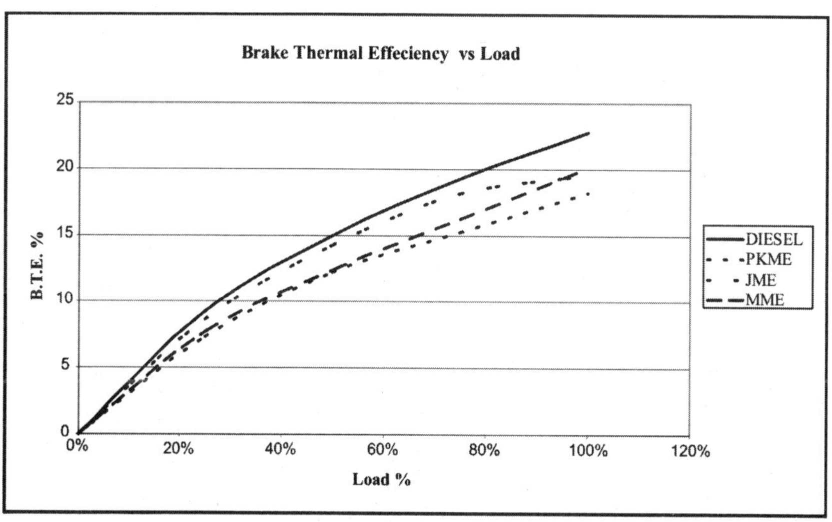

Fig. 5.16. Comparison of brake thermal efficiency Vs load per cent

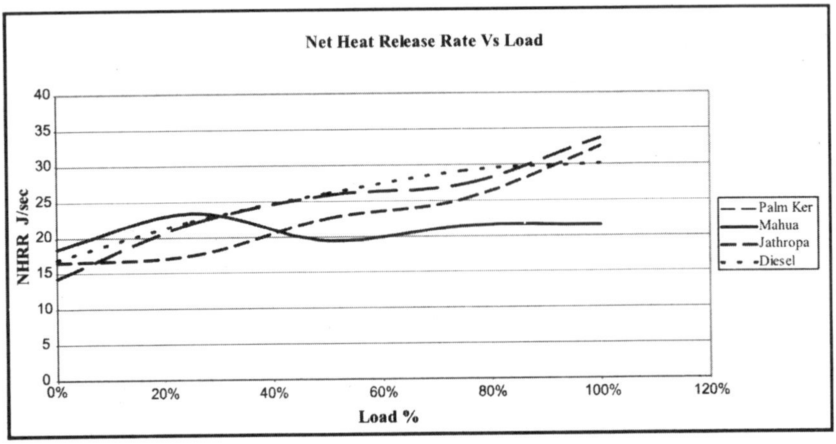

Fig. 5.17. Comparison of net heat release rate Vs load percentage

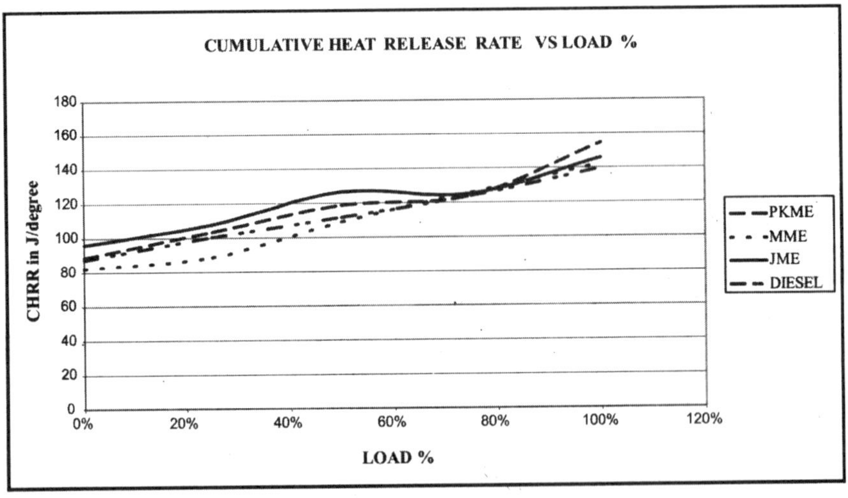

Fig. 5.18. Comparison of cumulative heat release rate Vs load percentage

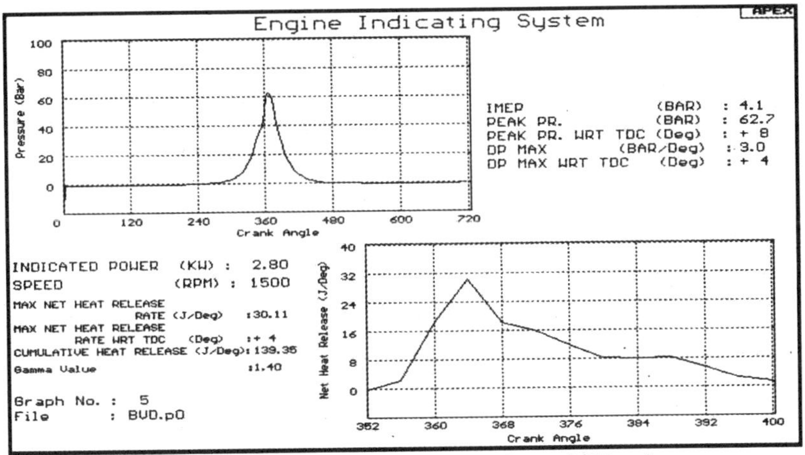

Fig. 5.19. Diesel heat release rate plot at full load

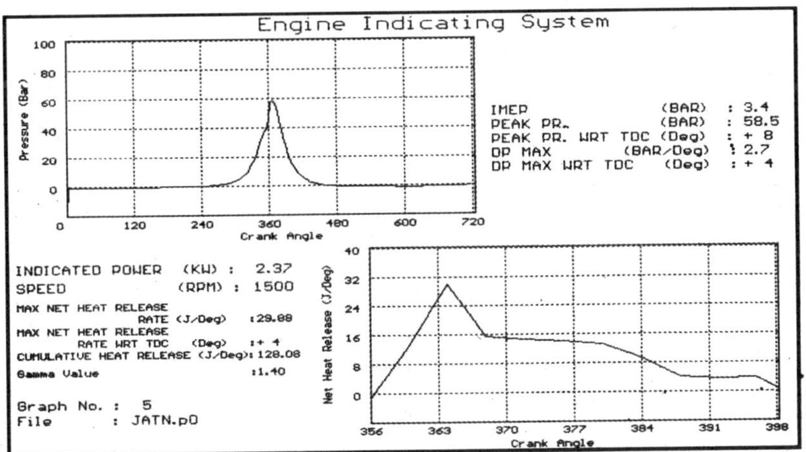

Fig. 5.20. JME heat release rate plot at full load

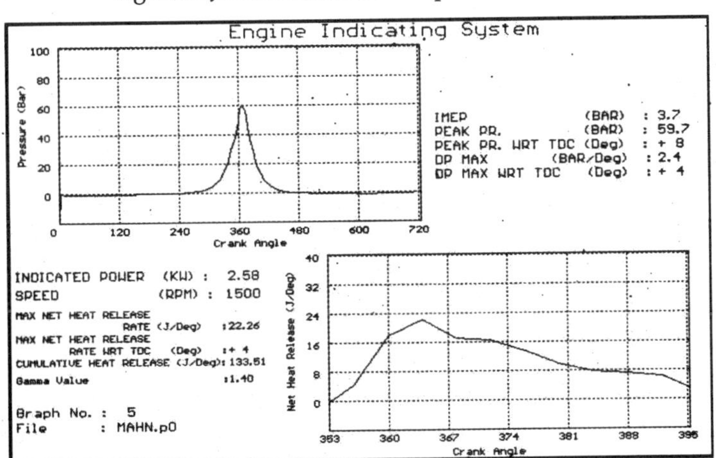

Fig. 5.21. MME heat release rate plot at full load

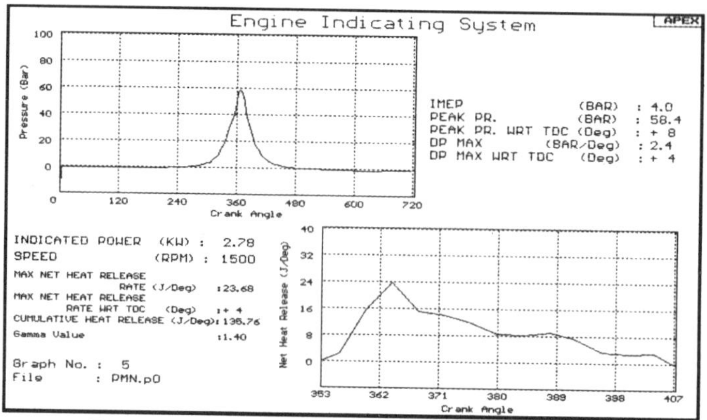

Fig. 5.22. PKME heat release rate plot at full load

COMPARISON OF THE SOUND PRESSURE LEVELS AT ALL LOADS AND FOR ALL THE OILS

Diesel oil produces more sound pressure levels than any other oil at all loads by a maximum rise of 1.5 decibels. The MME generates least noise (Fig. 5.23). It can be observed that the noise emanation from the cylinder depends on the indicated power developed with particular usage of the oil. Since the power developed by the MME is comparatively lesser, the sound pressure levels will also be lesser. Some investigations [35,36] in the literature revealed that the noise emanation depends on the maximum differential pressure rise and rate of pressure rise during combustion. The Figure 5.6 envisages the differential pressures that follow suit of noise of the engine run with the oils under investigation.

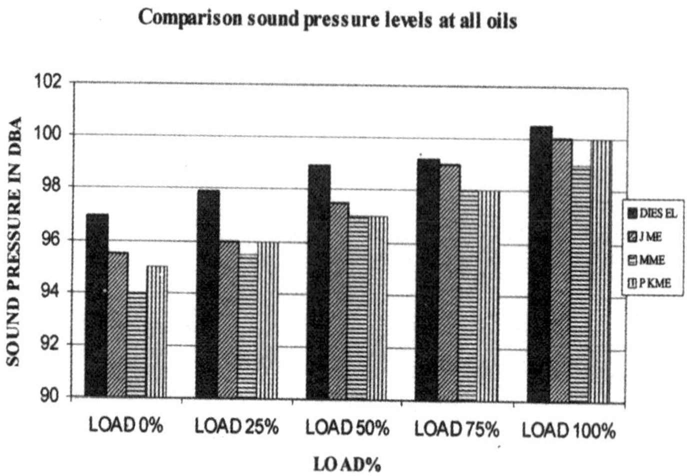

Fig. 5.23. Comparison of sound pressure levels recorded at various loads and for various oils

5.3 Engine Vibration Comparison

The overall values of vibration on the cylinder in two directions are measured. One in the radial direction perpendicular to the crankshaft and the second on the cylinder head in vertical direction.

With JME, the engine generated the minimum overall vibration when compared to other oils including diesel but with an exception that it generates slightly higher value on the cylinder head in vertical direction as shown in Figure 5.24

The vibration transmitted to the foundation in the case of diesel oil run is comparatively more than other oils as obtained in the vibration spectrums (Figs. 5.26-5.29). The vibration signatures obtained on the cylinder in vertical direction indicate lesser high frequency amplitudes in the case of JME (Fig. 5.31) indicating supporting lesser overall levels as indicated above. The observation is same for JME in the signatures obtained in the other two radial directions on the cylinder (Figs. 5.35 & 5.39).

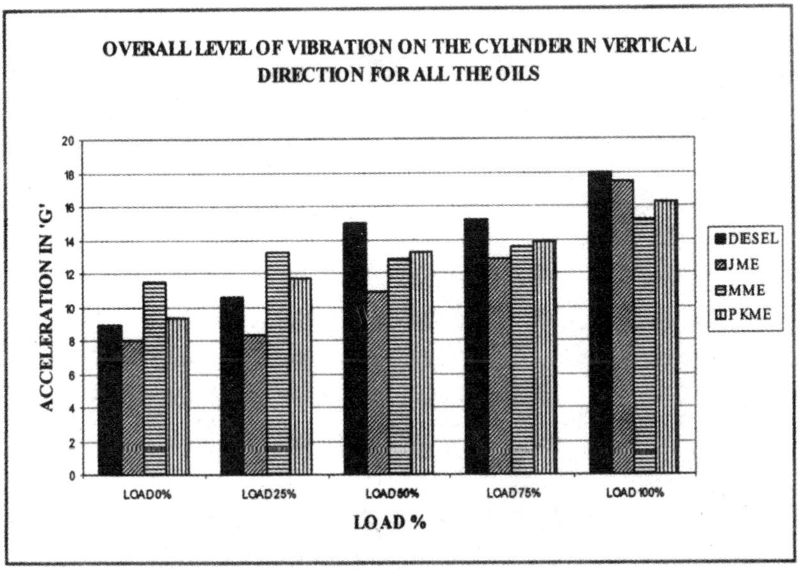

Fig. 5.24 .Overall values of vibration in acceleration amplitude Vs the load percentage

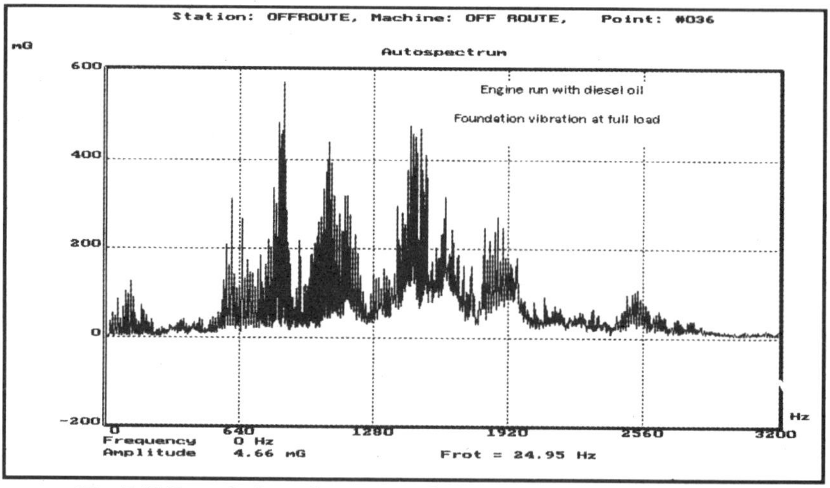

Fig. 5.25.Foundation vibration with the diesel oil run at full load

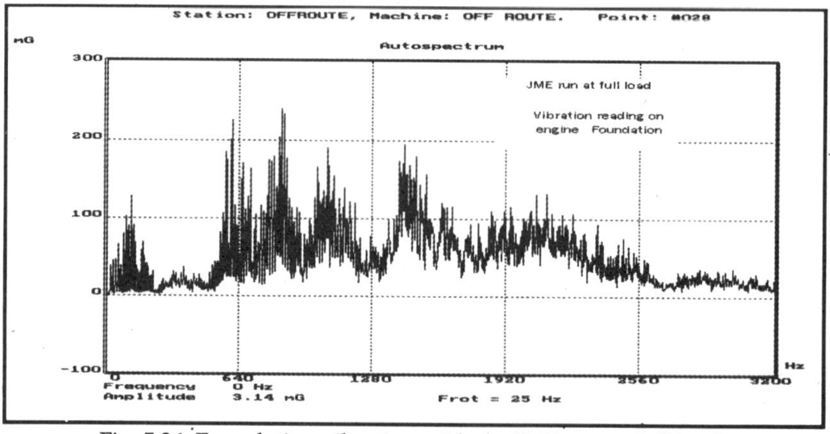

Fig. 5.26. Foundation vibration with the JME run at full load

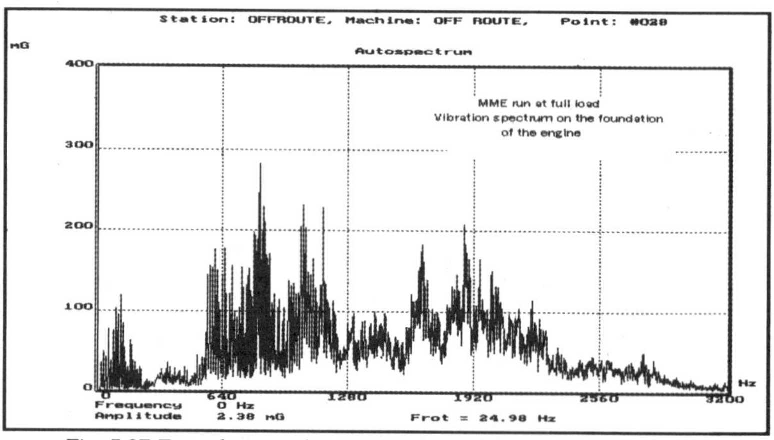

Fig. 5.27. Foundation vibration with the MME run at full load

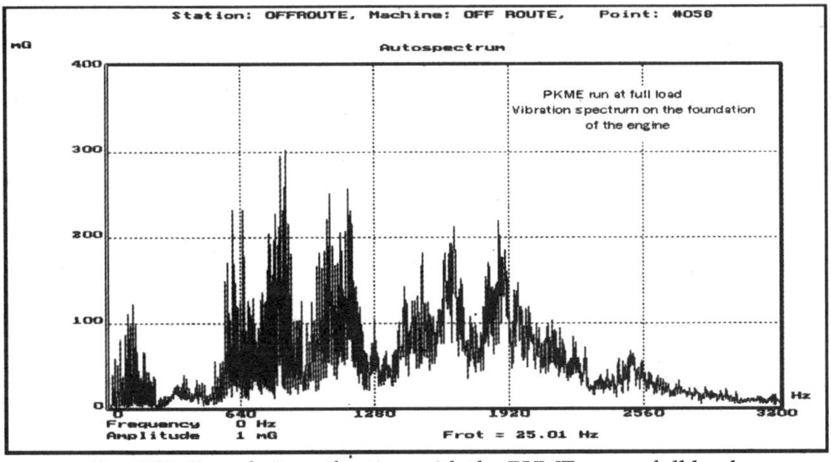

Fig. 5.28. Foundation vibration with the PKME run at full load

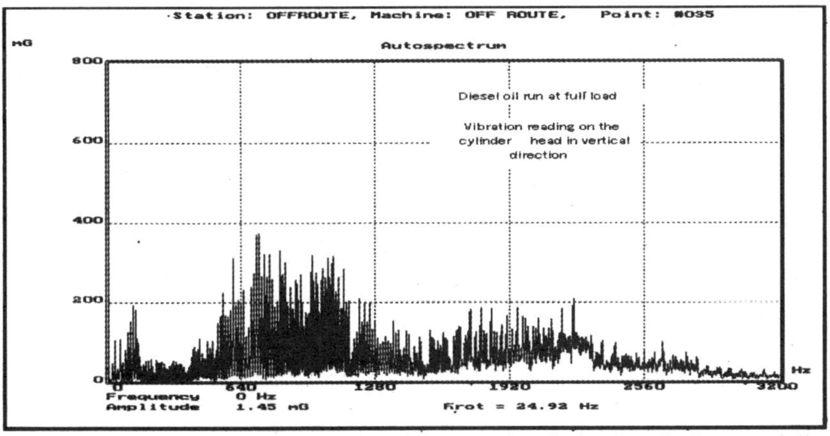

Fig. 5.29. Cylinder vibration in vertical direction with the diesel oil run at full load

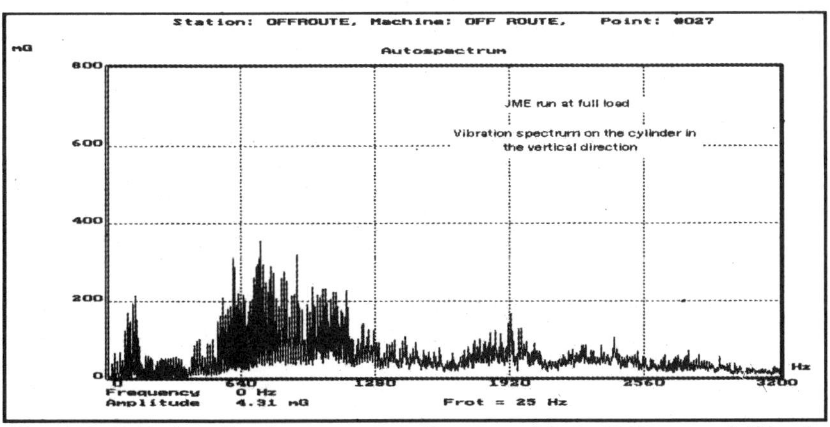

Fig. 5.30. Cylinder vibration in vertical direction with the JME run at full load

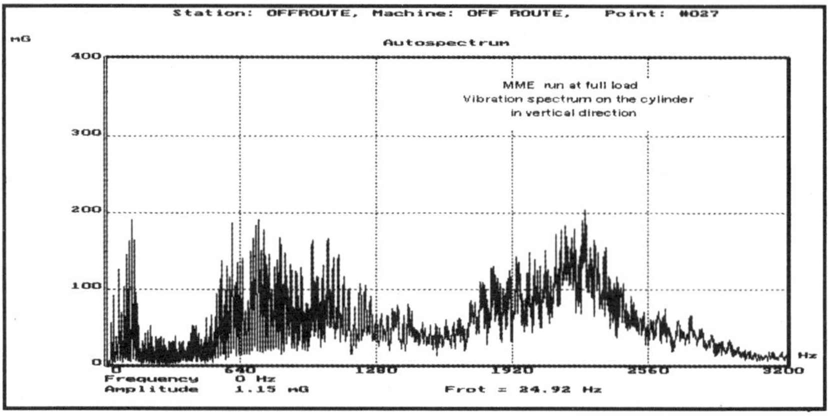

Fig. 5.31. Cylinder vibration in vertical direction with the MME run at full load

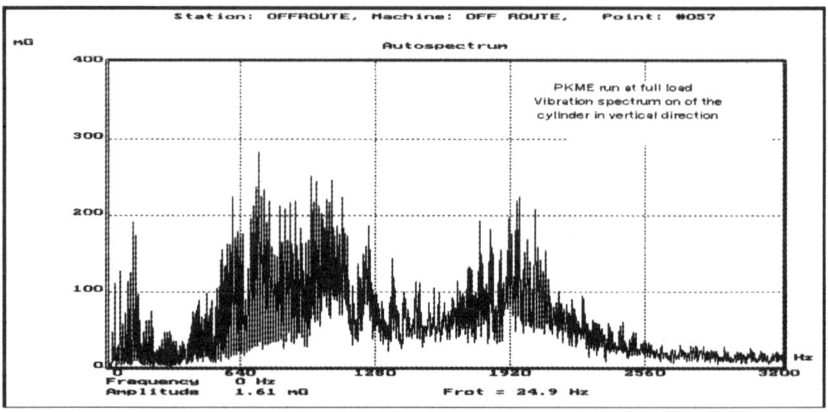

Fig. 5.32. Cylinder vibration in vertical direction with the PKME run at full load

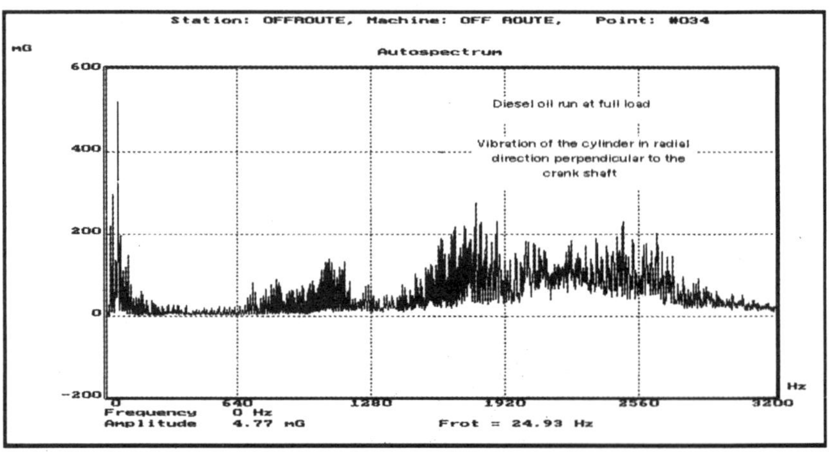

Fig. 5.33. Cylinder vibration in radial direction perpendicular to the crankshaft with the diesel oil run at full load

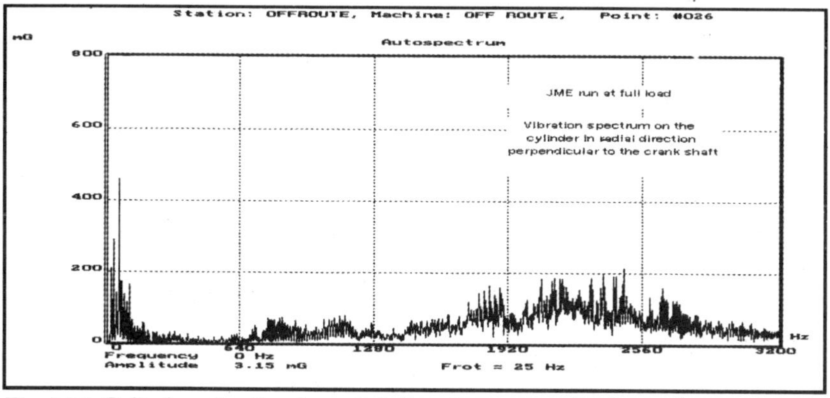

Fig. 5.34 .Cylinder vibration in radial direction perpendicular to the crankshaft with the JME run at full load

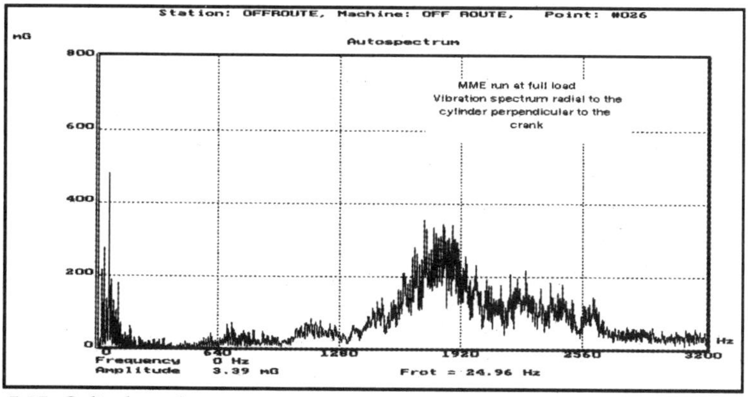

Fig. 5.35 .Cylinder vibration in radial direction perpendicular to the crankshaft with the MME run at full load

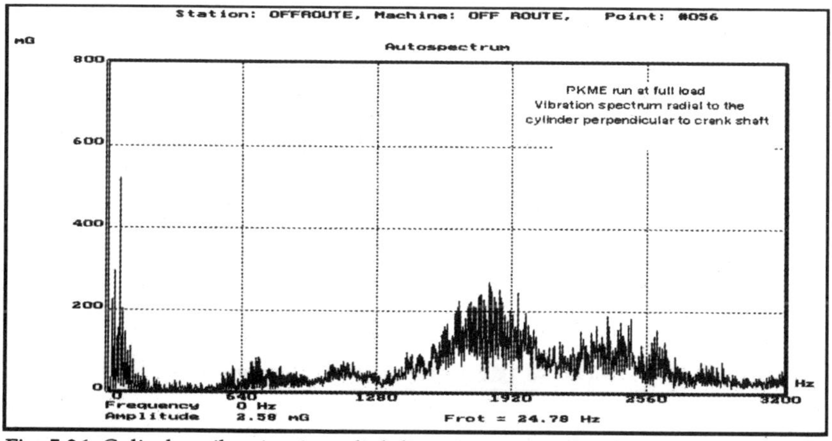

Fig. 5.36 .Cylinder vibration in radial direction perpendicular to the crankshaft with the PKME run at full load

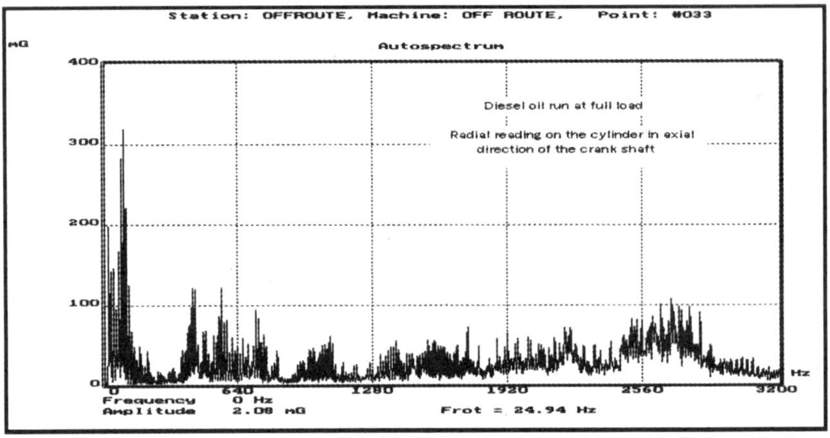

Fig .5.37. Cylinder vibration in radial direction axial to the crankshaft with the diesel oil run at full load

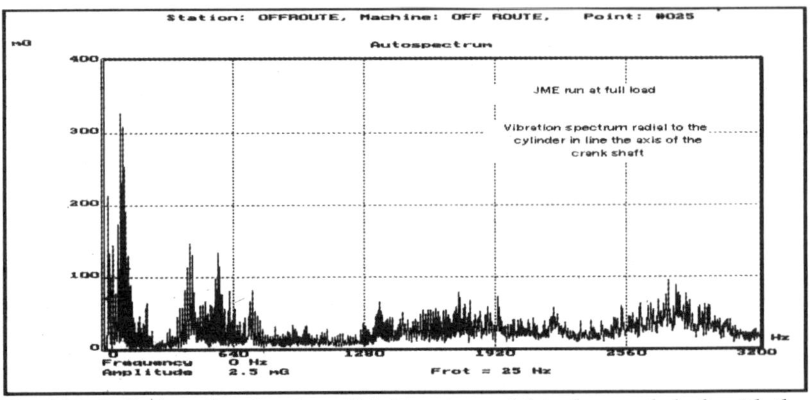

Fig. 5.38 .Cylinder vibration in radial direction axial to the crankshaft with the JME run at full load

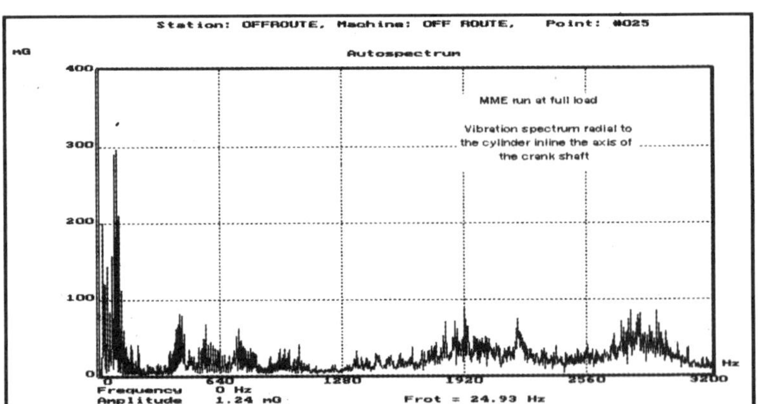

Fig. 5.39.Cylinder vibration in radial direction axial to the crankshaft with the MME run at full load

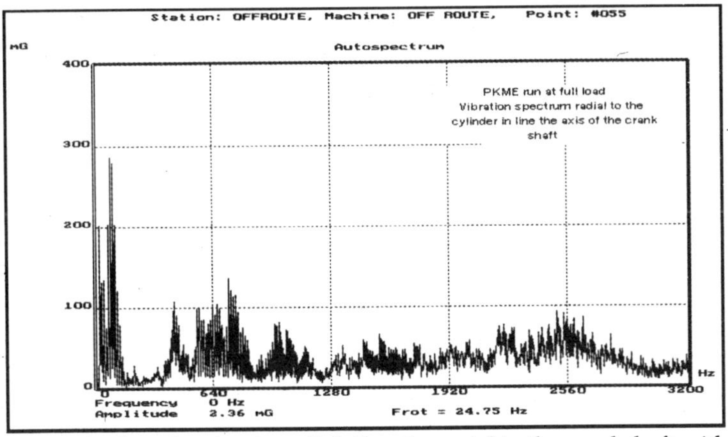

Fig. 5.40. Cylinder vibration in radial direction axial to the crankshaft with the PKME run at full load

Time Waveforms of Vibration

The time waveforms collected for comparison do not yield information as to identify the combustion propensities with the usage of oils under consideration. Combustion is identified to take place after every 80 ms. The Time waveforms (Figs. 5.41-5.48) are taken radially and vertically on the cylinder with the engine run with the methyl esters and the diesel oil. The time waveforms in the piston slap direction (Fig. 5.45-5.48) envisage higher vibration acceleration levels in case diesel oil burning approaching 5g and the acceleration levels in case of esters are below 4g. This is because of the lower power development in the case of the esters tested.

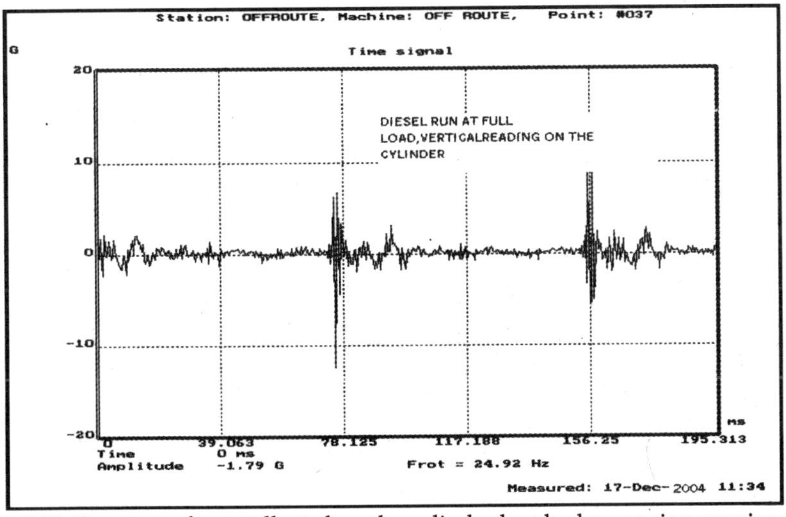

Fig. 5.41. Time waveform collected on the cylinder head when engine running at full load with the diesel oil

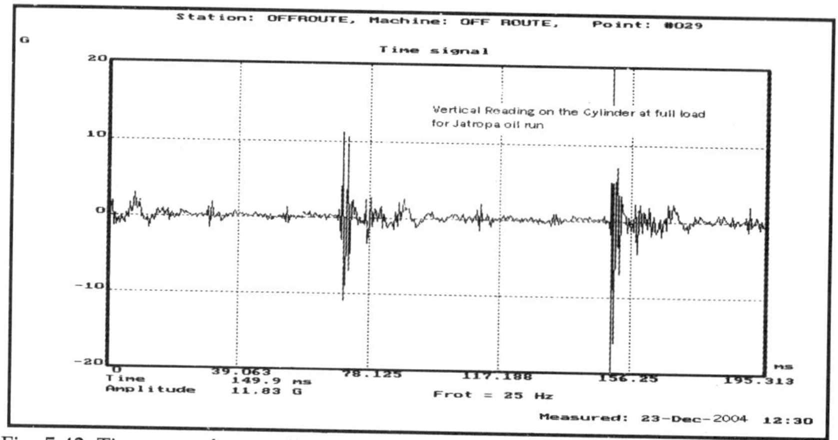

Fig. 5.42. Time waveform collected on the cylinder head when engine running at full load with the JME

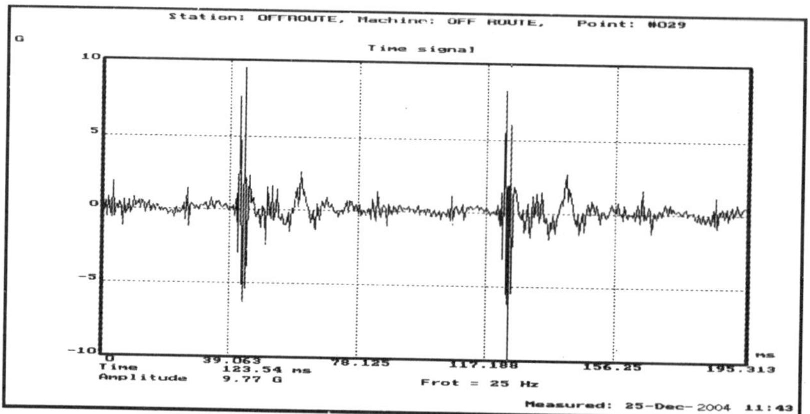

Fig. 5.43 .Time waveform collected on the cylinder head when engine running at full load with MME

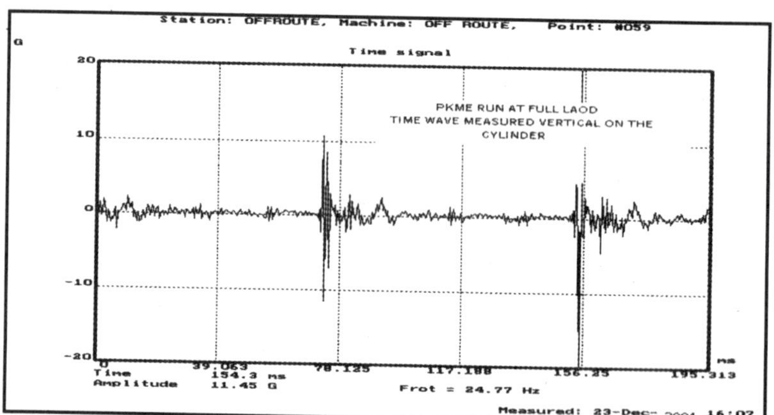

Fig. 5.44. Time waveform collected on the cylinder head when engine running at full load with PKME

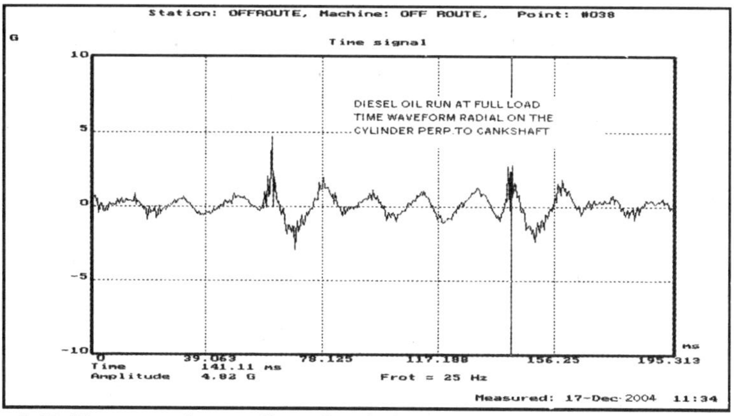

Fig. 5.45. Time waveform collected radial on the cylinder head when engine running at full load with the diesel oil.

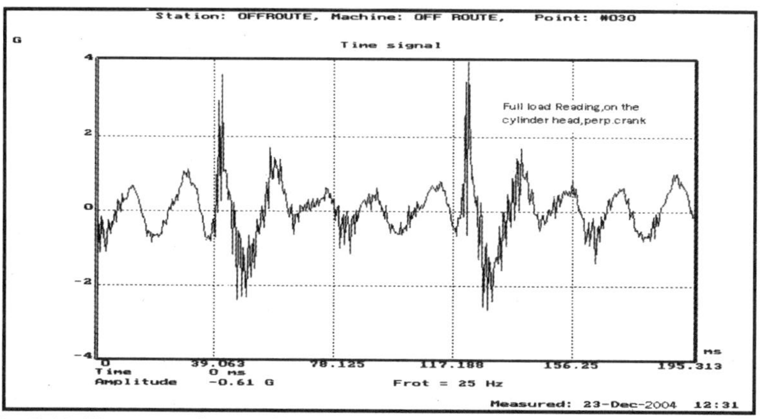

Fig. 5.46. Time waveform collected radial on the cylinder head when engine running at full load with JME

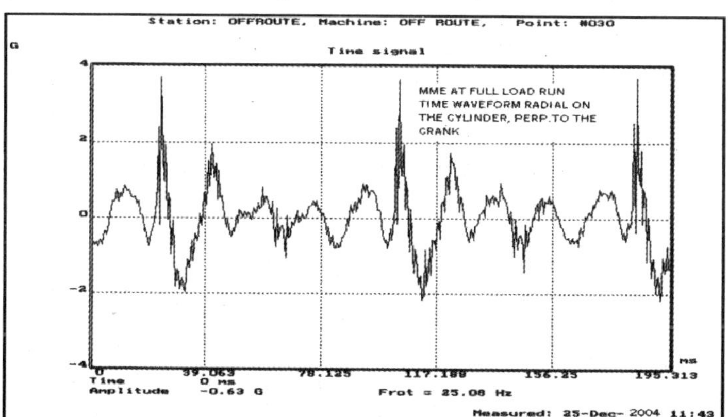

Fig .5.47. Time waveform collected radial on the cylinder head when engine running at full load with MME

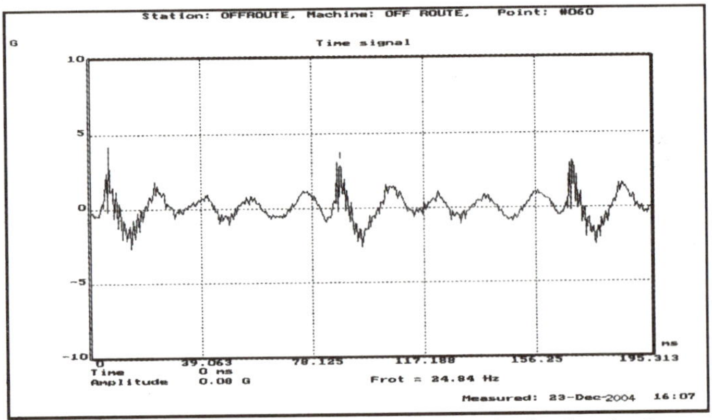

Fig. 5.48. Time waveform collected radial on the cylinder head when engine running at full load with PKME

Phase Analysis

The phase of vibration is an important measure to identify the mode of vibration concerning the order. The first order phase measurements are taken at two strategic positions on the cylinder viz. vertical and radial for the engine run with the esters and diesel oil (Fig. 5.49 & 5.50). The phase of vibration for the MME is different in the radial direction to the cylinder as observed in Figure 5.49. The phases measured vertical on the cylinder as shown in Figure 5.50 are different for the esters when compared to the diesel. This indicates change in mode of vibration for the first order. This may cause undue increase in the amplitude of vibration at other non-synchronous frequencies. This phase difference may affect the lower frequencies to generate higher amplitudes of vibration creating an impression that the combustion with the new oils is different. But the vibration signatures indicate that the esters tested gave reduced vibration levels when compared to the diesel oil. Hence a conclusion can be drawn that the effect of phase change is the representative of the reduced delay period, which can be observed from the Figures 5.19-5.22. The phase diagrams in the case of MME indicate early start of combustion.

Phase Of Vibration In *Degrees* (First Order) Radial Perp.To Crank On The Cylinder				
	Diesel	JME	MME	PKME
Load 0%	9.6	4	353.1	7
Load 25%	3	2.9	354.9	9
Load 50%	3	3	355.5	8.1
Load 75%	3.9	6.7	356.3	9.9
Load 100%	5.3	7.3	359.7	13

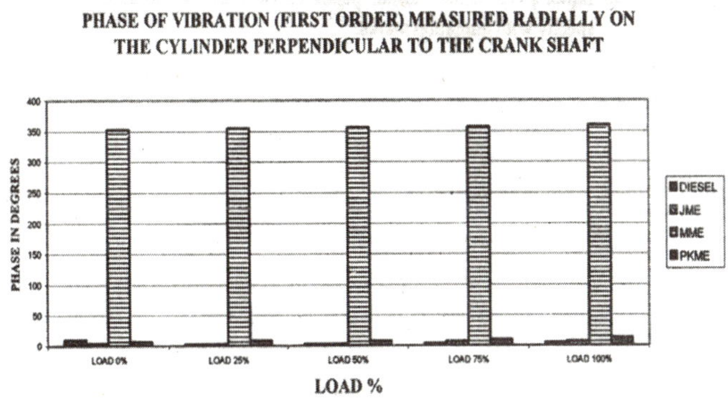

Fig. 5.49. Comparison of first order phase of vibration measured radial to the cylinder

Phase Of Vibration In *Degrees* (First Order) Vertical On The Cylinder				
	Diesel	JME	MME	PKME
Load 0%	6.3	335.8	333.1	343.4
Load 25%	31.1	352.5	339.4	344
Load 50%	21.8	2.1	342.5	345.4
Load 75%	5.5	338.4	336.4	1.8
Load 100%	35.9	340.1	338.7	355.4

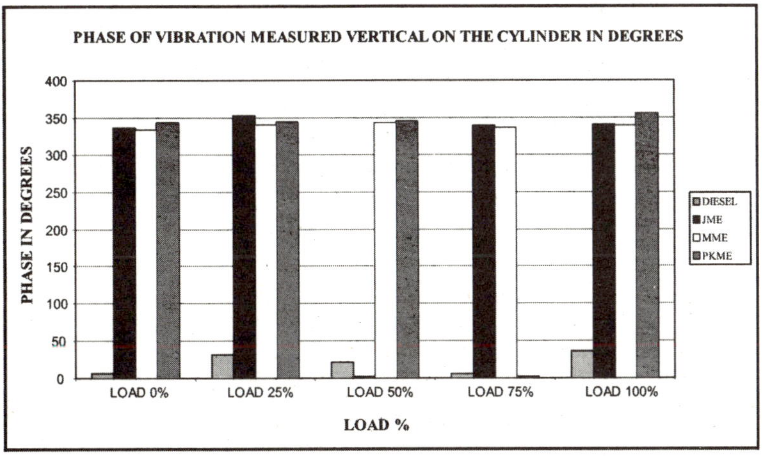

Fig. 5.50. Comparison of first order phase of vibration measured vertical on the cylinder

Assessment of Engine Trend with the Use of PKME, JME, MME and Diesel Oils

The data collected indicates that the vibration trend of the engine with the usage of diesel oil is severe as observed in the Figure 5.51 given by the On-Time software for vibration analysis. There is a sharp jump in overall value of vibration in the piston slap direction as per number '2' line from Jatropha to the diesel. Same increase holds good for the foundation vibration also. The overall level generated by the MME further increases in the direction mentioned above. Except this unexpected behaviour of MME, diesel predominates in vibration in all directions on the cylinder and at the foundation.

Date-wise succession of experimentation with the oils is as mentioned: PKME, JME, DIESEL, MME.

Trend curves are demarcated by colours from upper one to bottom one with the vibration readings and curve designations are as follows:

1. Vertical on the cylinder

2. Radial on the cylinder perpendicular to crank shaft

3. Foundation

4. Radial on the cylinder in line crankshaft

The Figure 5.52 depicts the vibration acceleration levels (spectrum averages) in the limited spectrum range of 3200 Hz with respect to the point of measurement on the engine test rig. Whereas earlier plots indicate the vibration levels based on the total spectrum averages at various loads. It can be observed that the vibration transmitted to the foundation is minimum and almost same for all oils as shown in the Figure. The amplitudes of vibrations in vertical and radial directions are complimentary for individual oils.

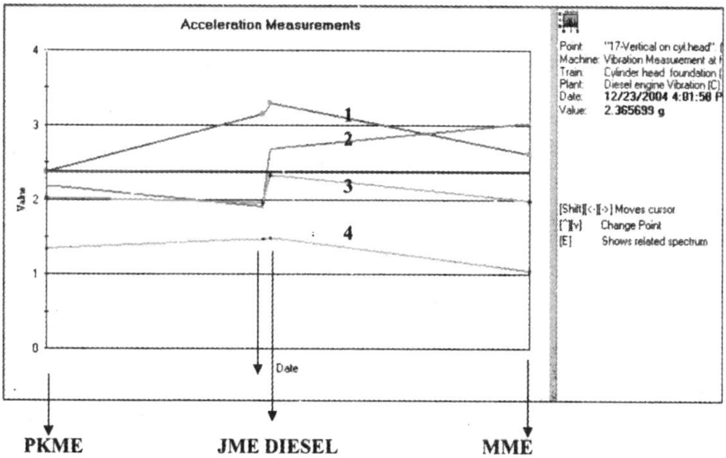

Fig. 5.51. Engine vibration trend when run with oils under consideration at full load

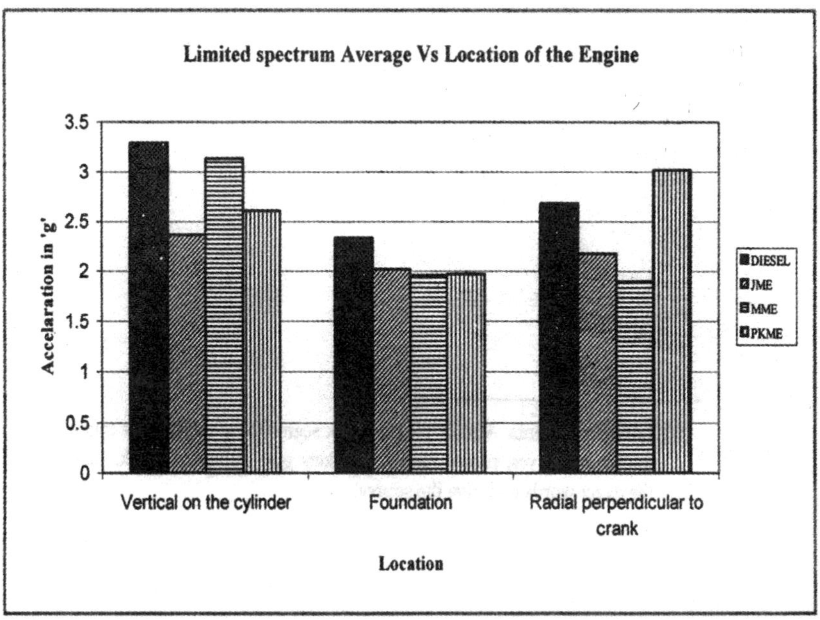

Fig. 5.52. Limited spectrum average of vibration acceleration when run with oils under consideration at full load

5.7 CONCLUSIONS

1. Fuel consumption in the case of diesel run is comparatively lower than any other ester tested on the same engine. Except in the case of mechanical efficiency, diesel excelled in almost all parameters for

which graphic evaluation is made. The reason can be assigned to the structure of the esters and oil lubricity because of which the frictional wastage of power might have been reduced.

2. The brake specific fuel consumption is an important parameter that is rated whenever engine performance is evaluated. In our observation, diesel oil's BSFC is comparatively lowest in comparison. The reason can be assigned to the inferior calorific values of the esters under consideration.

3. There is an increased vibrational severity of the engine with the usage of the diesel oil when compared to the esters tested. This is due to relatively higher power development in the case of diesel oil.

4. The noise emanation from the engine with the usage of diesel oil is more at all loads when compared to the esters. JME succeeds followed by PKME in the decreasing order of noise levels. This trend basically depends on the power generation in the cylinder by the usage of the independent oils. However, the aspects of knock and detonation for all oils are conspicuously absent as can be assessed from the P-θ plots.

5. Vibration phase measurements indicate that there is a phase difference for the first order which exceeds even 55^0. This will give an additional support to the conclusion that the diesel oil gives higher vibratory trend than other methyl esters and also the phase difference in the first order is indicative of the ignition delay.

6. In the light of the above said conclusions, even though the power development suffers to some extent, the esters can be recommended to be implemented to run the engine without changing the basic design of the engine because of the reason that the esters are affordable and renewable and also in view of on-going trends in the increase of prices of petro-diesel. The vibration point gives strength for the implementation in lieu of the petro-diesel.

7. **Cost Audit**: Edible oils are costlier than the non-edible oils. The non-edible oils are abundantly available in Indian forests. Some of the ethnic oils like Mahua are available in our forests and their usage are limited. It is a known fact that the palm kernel and Jatropha oils are available throughout the world. The cost of procurement and production of esters of these oils is comparatively less per

litre. Additionally, the by-product glycerine that is obtained from the esterification process is costlier and has more market value.

Future Scope of Work

- As a part of future scope of work, the pollutants like NO_X, CO and unburnt hydrocarbons can be studied.
- The same methyl esters can be tested using variable C.I. engine in order to fix the most useful compression ratio.
- Adiabatic engines can be tested with these methyl esters and the performance can be evaluated.

Bibliography

Agarwal, A. K. "Performance Evaluation and Emission Characteristics of a Compression Ignition Engine Using Esterified Biodiesel, M Tech thesis." *Centre for Energy Studies, Indian Institute of Technology*, 1992: pg. no. 82 First citation in article.

Ali, Y., and M. A. Hanna. "Alternative diesel fuels from vegetable oils." *Bioresource technology : biomass, bioenergy, biowastes, conversion technologies, biotransformations, production technologies* Vol. 50, no. 2, (1994): Pages 153-163 First citation in article.

Ali, Y., K. M. Eskridge, and M. A. Hanna. ""Testing of alternative diesel fuel from tallow and soybean oil in Cummins N14-410 diesel engine"." *Bioresource Technology*, 1995a: Vol. 53 :Pages 243-254.

Ali, Yusuf, and Milford A. Hanna. *In-cylinder pressure characteristics of a D.I. heavy duty diesel engine on biodiesel fuel.* Warrendale, Pa: Society of Automotive Engineers, 1997 Paper no. 971683.

Andrzejewski, J., and A. Sapinski. "Some Particularities of the Vegetable Oils Combustion in Diesel Engines." *Conference: Engine and Environment.* AVL Proceedings, 1991.

Babu, A. K., and G. Devaradjana. *Vegetable oils and their derivatives as fuels for CI engines : an overview.* Madras: Anna University, Department of Automobile Engineering, 2003 Paper no. 2003-01-0767.

Barbella, R. *In-cylinder sampling of high molecular weight hydrocarbons from a D.I. light duty diesel engine.* Warrendale, PA: Society of Automotive Engineers, 1989 Paper no. 89-0437.

Barsic, N. J., and A. L. Hunke. *Performance and emission characteristics of a naturally aspirated diesel engine with vegetable oils.* Society of Automotive Engineers, 1981 Paper no. 810262.

Bechtold, Richard L. *Alternative fuels guidebook : properties, storage, dispensing, and vehicle facility modifications.* Warrendale, Pa: Society of Automotive Engineers, 1997.

Cardone, M., M. V. Prati, V. Rocco, and A. Senatore. *Analisi Sperimentale delle Prestazioni e delle Emissioni di un Motore Diesel Alimentato con Miscele di Biocombustibile e Gasolio.* Roma, Febbraio: MIS-MAC V Metodi di Sperimentazione nelle Macchine, 1998.

Chandler, Kevin. *Alternative fuel transit bus evaluation program results.* Warrendale, Pa: Society of Automotive Engineers, 1996 Paper no. 961082.

Choi, C. Y., G. R. Bower, and R. D. Reitz. *Effects of biodiesel blended fuels and multiple injections on D. I. diesel engines.* Warrendale, Pa: Society of Automotive Engineers, 1997 Paper no. 970218.

Clark, S. J., L. Wanger, M. D. Schrock, and P. G. Piennaar. "Methyl and Ethyl Soybean Ester as renewable fuels for diesel engines." *Journal of American Oil Chemists Society*, 1984: Vol. 61 (1984) (10), pp. 1632–1637.

Culshaw, F. *The potential of biodiesel, from oilseed rape.* London, England: Institution of Mechanical Engineers, 1993.

Czerwinski, J. *Performance of HD-DI-diesel engine with addition of ethanol and rapeseed oil.* Warrendale, Pa: Society of Automotive Engineers, 1994.

Desantes, Jose M., Jean Arregle, and Santiago Ruiz. *Characterisation of the injection-combustion process in a D.I. diesel engine running with rape oil methyl ester.* Warrendale, Pa: Society of Automotive Engineers, 1999.

Dulger, Zafer. *Utilization of Sunflower Methyl Ester as a Diesel Engine Fuel.* Society of Automotive Engineers, 2001 Paper no. 2001-01-3633.

Freedman, B., and M. O. Bagby. *Diesel Engine Fuels from Vegetable Oils.* Cincinnati, Ohio: American Oil Chemists Society, 1989.

Freedman, B., E.H Pryde, and Mounts T.L.J. American Oil Chemistry Society , 61, 1638, 1984.

Freedman, B., R. O. Butterfield, and E.H. J. Pryde. American Oil chemistry Society 63, 1375, 1986.

Garret, T. K. *Automotive fuels and fuel systems : fuels, tanks, delivery, metering, mixing and combustion, and environmental considerations.* Warrendale, PA: Society of Automotive Engineers, 1991.

Gatowski, J. A. *Heat release analysis of engine pressure data.* Warrendale, PA: Society of Automotive Engineers, 1984 Paper no. 841359.

Gerpen, Jon H. Van. *Determining the influence of contaminants on biodiesel properties.* Warrendale, Pa: Society of Automotive Engineers, 1997 Paper no. 971685.

Gill, Alan P. *Design choices for 1990's low emission diesel engines.* Warrendale, PA: Society of Automotive Engineers, 1988 Paper no. 88-0350.

Gopalkrishnan, K. V., and P. S. Rao. *Use of non-edible vegetable oils as alternate fuels in diesel engines. DNES Project report.* Madras-36: ICE Lab IIT, 1996.

Gunter, H. *Rape Oil Methyl Ester (RME) and Used cooking Oil Methyl Ester (UOME) as Alternative Fuels.* Society of Automotive Engineers, 1995 Paper no. 952755.

Hayes, T K. *The effect of fumigation of different ethanol proofs on a turbocharged diesel engine.* Warrendale, PA: Society of Automotive Engineers, 1988.

Howell, Steve. *U. S. biodiesel standards : an update of current activities.* Warrendale, Pa: Society of Automotive Engineers, 1997 Paper no. 971687.

Kaiser W. "The Use of Alternative Fuels in Tractor Engines- State of Development and Prospects from a Manufacturer's Viewpoint." *Conference : Engine and Environment.* AVL Proceedings, 1991.

Kalschmitt, M., G. A. Reinhardt, and T. Stelzer. "Life cycle analysis of biofuels under different environmental aspects." *Biomass & bioenergy* (Biomass & bioenergy), 1997: Vol. 12, no. 2, (1997): pages 121-134 First citation in article.

Karim, G. A., and N. Amoozegar. *Determination of the performance of a dual fuel diesel engine with the addition of various liquid fuels to the intake charge.* Warrendale, PA: Society of Automotive Engineers, 1983 Paper no. 830265.

Kaufman, K.R., Ziejewski, M.,. "Sunflower methyl esters for direct injection diesel engines." *Transactions of the American Society of Agricultural Engineers,* 1984: pp.1626-1633.

Kidoguchi, Y., C. Yang, and K. Miwa. "Effects of Fuel Properties on Combustion and Emission Characteristics of a Direct-Injection Diesel Engine." *S.A.E. transactions* Vol. 109, Part 4 (2000):Pages 1149-1157, Paper no. 2000-01-1851.

Kittelson, David B. *In-cylinder measurements of soot production in a direct-injection diesel engine.* Warrendale, PA: Society of Automotive Engineers, 1988 Paper no. 88-0344.

Klimstra, Jacob. *The optimum combustion phasing angle : a convenient engine tuning criterion.* Warrendale, PA: Society of Automotive Engineers, 1985 Paper no. 852090.

Knothe, Gerhard, Robert O. Dunn, and Marvin O. Bagby. "Biodiesel: The Use of Vegetable Oils and Their Derivatives as Alternative Diesel Fuels." *ACS symposium series* Vol. 666, (1997): Pages 172-208.

Koerbitz, W. "Biodiesel production in Europe and North America, an encouraging prospect." *Renewable energy* Vol. 16, no. 1/4, (1999): Pages 1078-1083 First citation in article.

Krieger, R. B., and G. Borman. *"The computation of Apparent Heat release for Internal combustion engines".* American Society of Mechanical Engineers, paper 66-WA/DGP-4, 1956.

Kumar, M. S., A. Ramesh, and B. Nagalingam. "Investigations on the use of Jatropha oil and its methyl ester as a fuel in a compression engine." *Journal of the Institute of Energy*. Vol. 498, (2001): Pages 24-28.

Lele, Satish. *Biodiesel and jatropha plantation.* Jodhpur, India : Agrobios, 2006.

Macchi S. "Overview on Biodiesel Utilisation for S+L+H Tractor Engines." *Conference : Engine and Environment.* AVL Proceedings, 1991.

Masjuki, H., M. Z. Abdulmuin, and H. S. Sii. "Investigations on preheated palm oil methyl esters in the diesel engine." *Proceedings of the Institution of Mechanical Engineers. Part A, Journal of power and energy* Vol.210, no. 2, (1996): page no. 131.

Mittelbach, M. "A European Perspective on Quality Biodiesel, Commercialization of Biodiesel: producing a quality fuel,." Edited by C. L. Peterson. *Conference Proceedings 9–10 July 1997, Boise, ID, University of Idaho, Moscow, ID, pp. 125–131, 1998.* First citation in article.

Moser, F. X. "Behaviour of Rapeseed Oil Methyl Ester Fuelled Tractor Engines and Field Experience." *Conference : Steyt-Daimler-Puch AG 3rd International Symposium.* The Performance Evolution of Autómotive Fuels and Lubricants, 1989.

Murayama, Tadashi. *Low carbon flower buildup, low smoke, and efficient diesel operation with vegetable oils by conversion to mono-esters and blending with diesel oil or alcohols.* Warrendale, PA: Society of Automotive Engineers, 1984 Paper no. 841161.

Niehaus, R. A., C. E. Goering, and L. D. Savage. "Cracked soybean oil as a fuel for a diesel engine." *Transactions of the American Society of Agricultural Engineers*, 1986: Vol. 29(3) :Pages 683-689.

Pachter, H., and G. Hohl. "Rapeseed Oil Methyl Ester (RME) as an Alternative Diesel Fuel Fleet Trials in the Austrian Army." *Conference : Engine and Environment.* AVL Proceedings, 1991.

Perkins, L. A., and C. L. Peterson. *Durability testing of transesterified winter rape oil (Brassica Napus L.) as fuel in small bore, multi-cylinder, DI, CI engines.* Warrendale, Pa: Society of Automotive Engineers, 1991 Paper no. 911764.

Peterson, C. L. "Vegetable oil as a diesel fuel." *Transactions of the American Society of Agricultural Engineers*, 1986: pp.1413-1422.

Pischinger, G., R. W. Siekmann, A. M. Falcon, and F. Fernandes. "Results of Engine and Vehicle Tests with Methyl Esters of Plant Oils as Alternative Diesel Fuels." *Proceedings of Fifth International conference on*

"Alcohol fuel technology,". Auckland, New Zealand, 1982 First citation in article.

R., Zilman. "Rapeseed Oil and Rapeseed Methyl Ester - Fuels for the Future." *International Conference : European Association of Cognitive Ergonomics.* 1993.

Rakapoulos, C. D. "Olive Oil as a Fuel Supplement in DI and IDI Diesel Engines." *Energy (Oxford)*, 1992: Vol. 17, pp. 787–790 First citation in article.

Ramesh, A., B. Nagalingam, and K. V. Gopalkrishnan. "A simple method to obtain the heat release rates of a diesel engine using a single analyzer." *National conference on I C engines and combustion.* Dec 17-19, 1987. PA-55.

Raneses, A. R., L. K. Glaser, J. M. Price, and J. A. Duffield. "Potential Biodiesel Markets and their Economic Effect on the Agricultural Sector of the United States." *Industrial Crops and Products,*, 1999: Vol. 6, pp. 151–162 First citation in article.

Rao, P. S., and K. V. Gopalkrishnan. "Vegetable Oils and Their Methyl Esters as Fuels for Diesel Engines." *Indian Institute of Technology, first citation in article*, 1991: Vol. 29, pp. 292–297.

Rickeard, D. J., and (N.D.) Thompson. *A review of the potential for bio-fuels as transortation fuels.* Warrendale, Pa: Society of Automotive Engineers, 1993 Paper no. 932778.

Rocco, V., M. V. Prati, A. Senatore, E. Cerri, and M. Benfenati. *Fuelling a Small Diesel Engine by Using Diesel Fuel and RME : Comparative Results and Particulate Characterization.* Lecce: 1o Convegno Internazionale sui Carburanti Alternativi e I Combustibili, 1996.

Schlick, M. L., M. A. Hanna, and J. L. Schinstock. *Soybean and sunflower oil performance in diesel engine.* American Society of Agricultural Engineers, 1988.

Scholl, Kyle W., and Spencer C. Sorenson. *Combustion of soybean oil methyl ester in a direct injection diesel engine.* Warrendale, Pa: Society of Automotive Engineers, 1993 Paper no. 930934.

Schuchardt, Ulf, Ricardo Sercheli, and Rogerio Atheus Vargas. *Transesterification of Vegetable Oils : A Review.* Brazil: Journal of the Brazilian Chemical Society, 1997.

Schumacher, L. G., and J. V. Gerpen. *Research needs resulting from experiences of fueling of diesel engines with biodiesel.* 1999 web.missouri.edu.

Senatore, A. *A Comparative analysis of combustion process in D.I. diesel engine fueled with biodiesel and diesel fuel.* Warrendale, PA: Society of Automotive Engineers, 2000 Paper no. 2000-01-0691.

Shundoh, Shigeru. *The effect of injection parameters and swirl on diesel combustion with high pressure fuel injection.* Warrendale, PA: Society of Automotive Engineers, 1991 Paper no. 91-0489.

V. Rocco, R. Montanari, M.V. Prati, A. Senatore. "Analysis of Particulate Emitted from an IDI Turbocharged Small Diesel Engine Fuelled with Diesel Fuel, Biofuel and their Blend." *Fall Technical Conference.* American Society of Mechanical Engineers, 1995. Volume 3.

Varaprasad, R. Rao. *Experimental investigation on the performance, Combustion and Emission Characteristics of an Alcohol Fuelled Single-cylinder S I Engine Ph. D. Thesis.* Chennai: Indian Institute of Technology, 1986.

Vellguth, Gerhard. *Performance of vegetable oils and their monoesters as fuels for diesel engines.* Warrendale, Pa: Society of Automotive Engineers, 1983 Paper no. 831358.

Venedaal, R., U. Joergensen, and C. A. Foster. "European energy crops: a synthesis." *Biomass & bioenergy* Vol.13, no. 3, (1997): Pages 147-185 First citation in article.

Zubik, J., S. Sorenson, and C. Georing. "Diesel engine combustion of Sunflower oil fuels." *Transactions of the American Society Agricultural Engineers,* 1984: pp.1252-1256.

Appendix A

The vibration signatures at other loads not mentioned in the chapters have been appended below for verification.

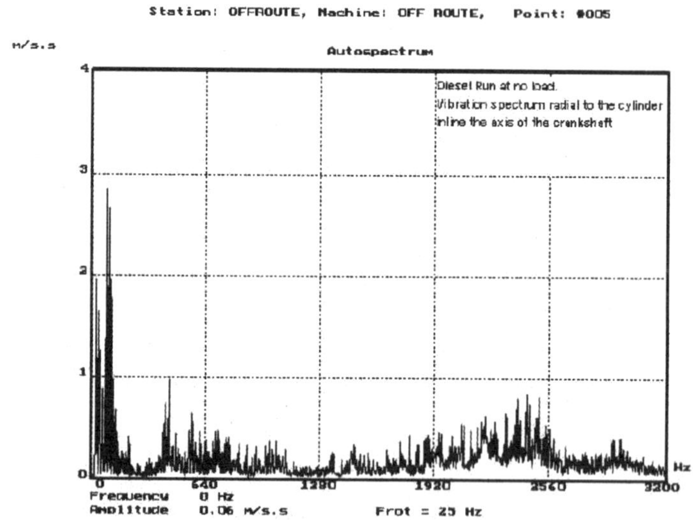

Fig:(A-1) Diesel run at no load. Vibration spectrum radial to the cylinder inline the axis of the crankshaft.

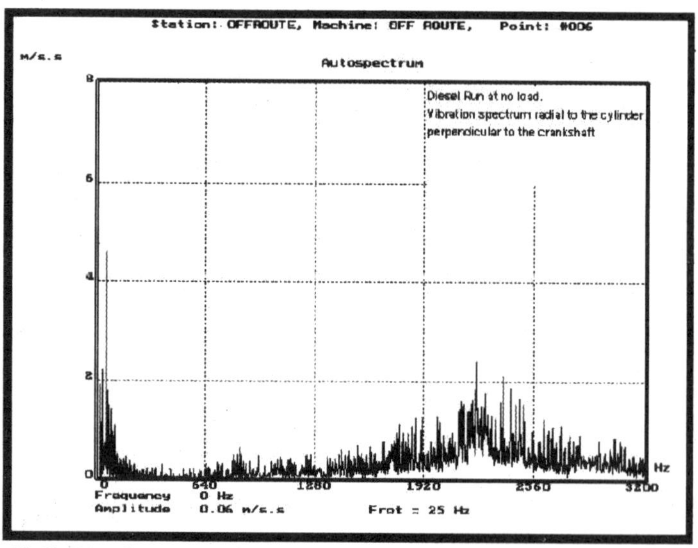

Fig:(A-2) Diesel run at no load. Vibration spectrum radial to the cylinder perpendicular to the crankshaft.

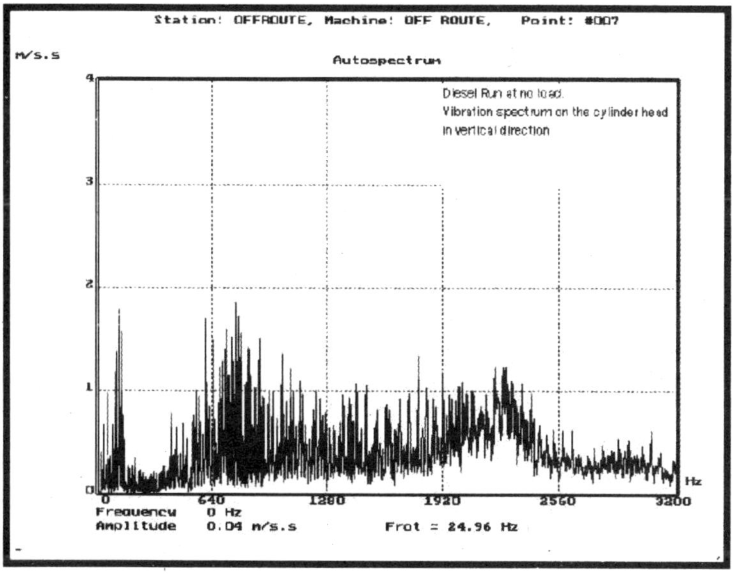

Fig: (A-3) Diesel run at no load. Vibration spectrum on the cylinder head in vertical direction.

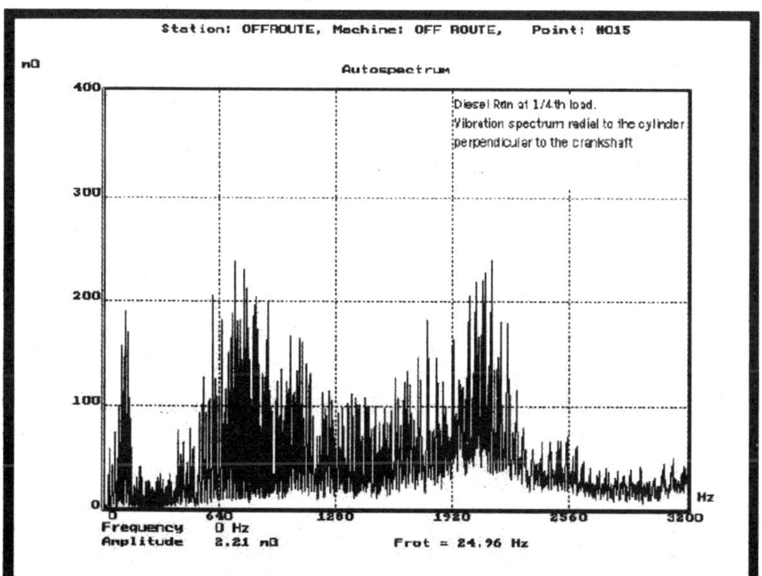

Fig: (A-4) Diesel run at no load. Vibration spectrum on the foundation of the engine.

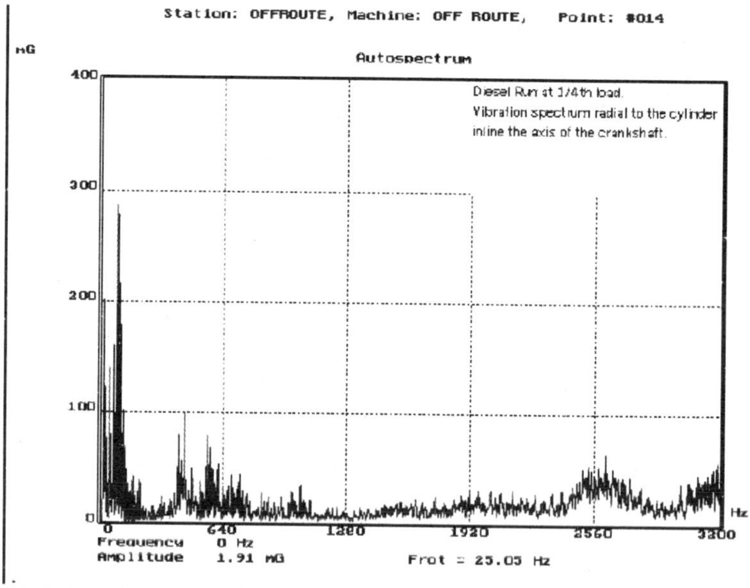

Fig:(A-5) Diesel run at 1/4th load. Vibration spectrum radial to the cylinder inline the axis of the crankshaft.

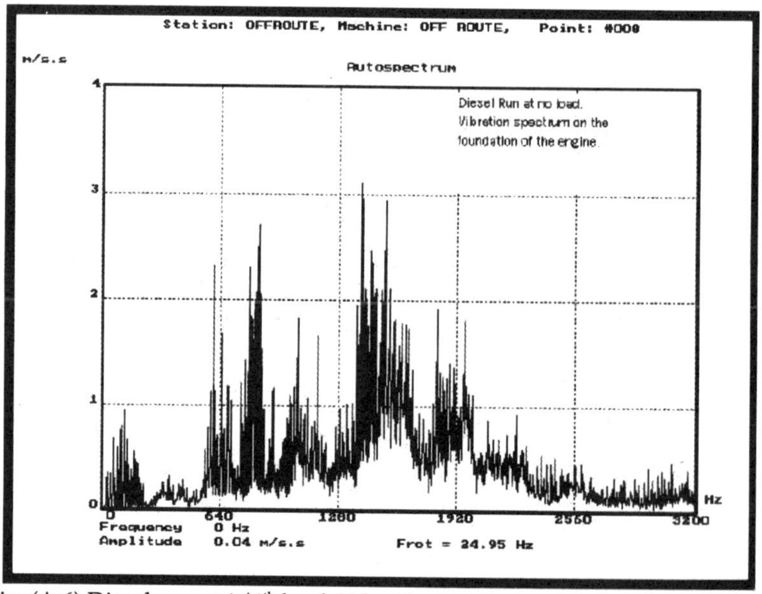

Fig: (A-6) Diesel run at 1/4th load. Vibration spectrum radial to the cylinder perpendicular to the crankshaft.

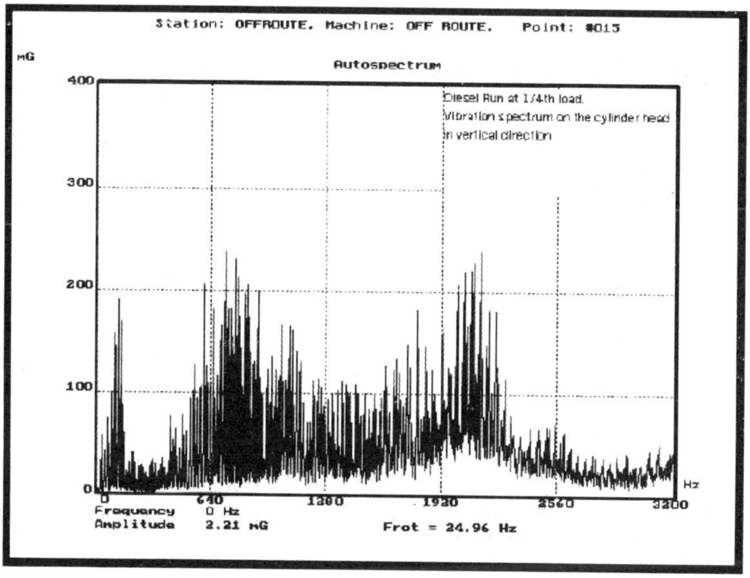

Fig:(A-7) Diesel run at 1/4th load. Vibration spectrum on the cylinder head in vertical direction.

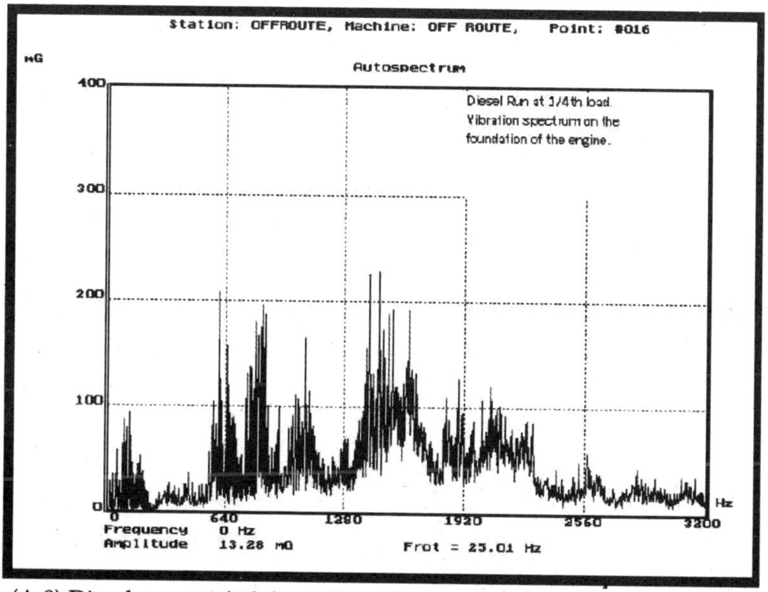

Fig: (A-8) Diesel run at 1/4th load. Vibration spectrum on the foundation of the engine.

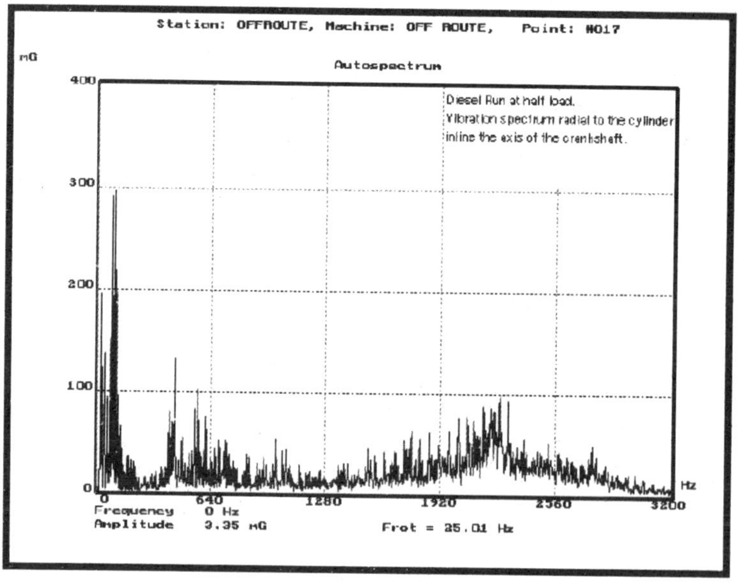

Fig:(A-9) Diesel run at half load. Vibration spectrum radial to the cylinder inline the axis of the crankshaft.

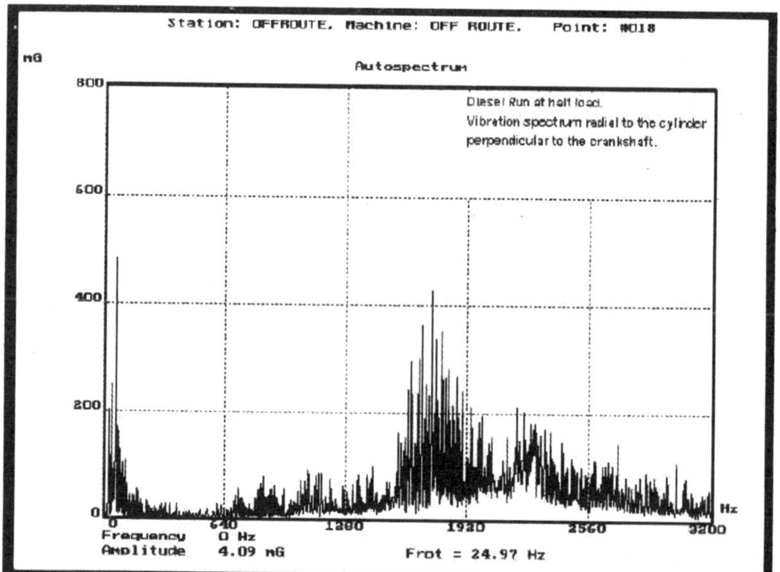

Fig: (A-10) Diesel run at half load. Vibration spectrum radial to the cylinder perpendicular to the crankshaft.

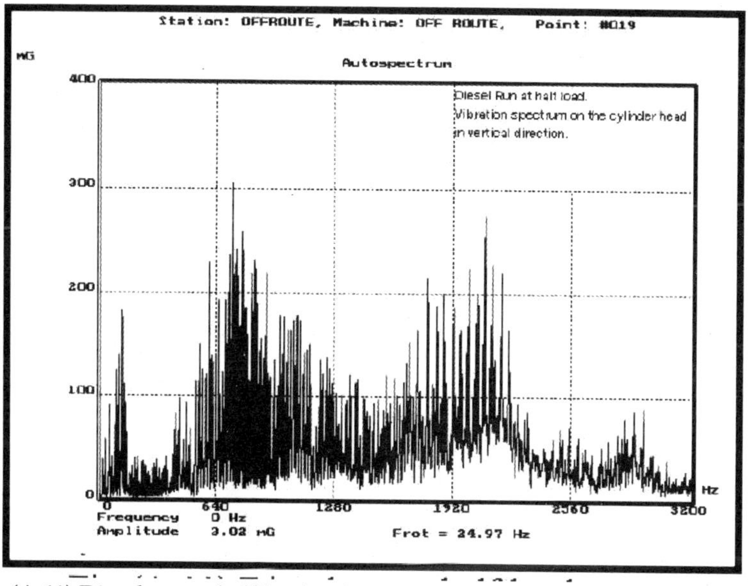

Fig: (A-11) Diesel run at half load. Vibration spectrum on the cylinder head in vertical direction.

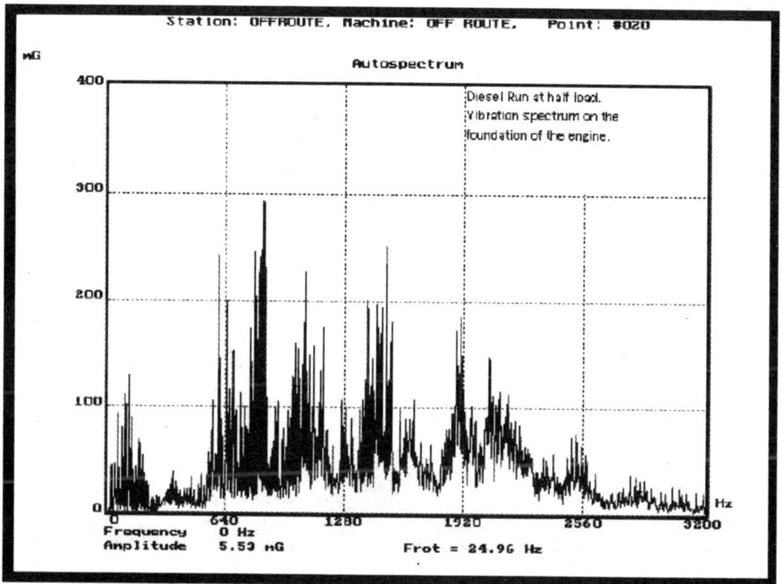

Fig:(A-12) Diesel run at half load Vibration spectrum on the foundation of the engine.

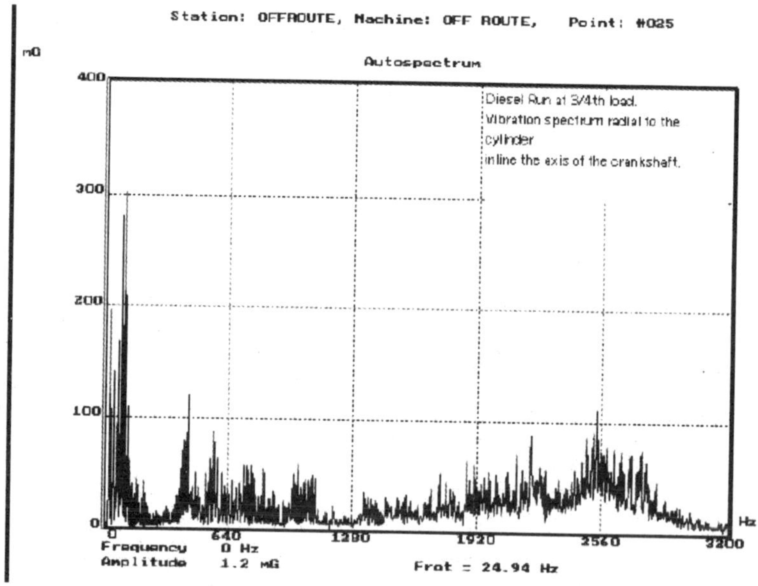

Fig:(A-13) Diesel run at 3/4th load. Vibration spectrum radial to the cylinder inline the axis of the crankshaft.

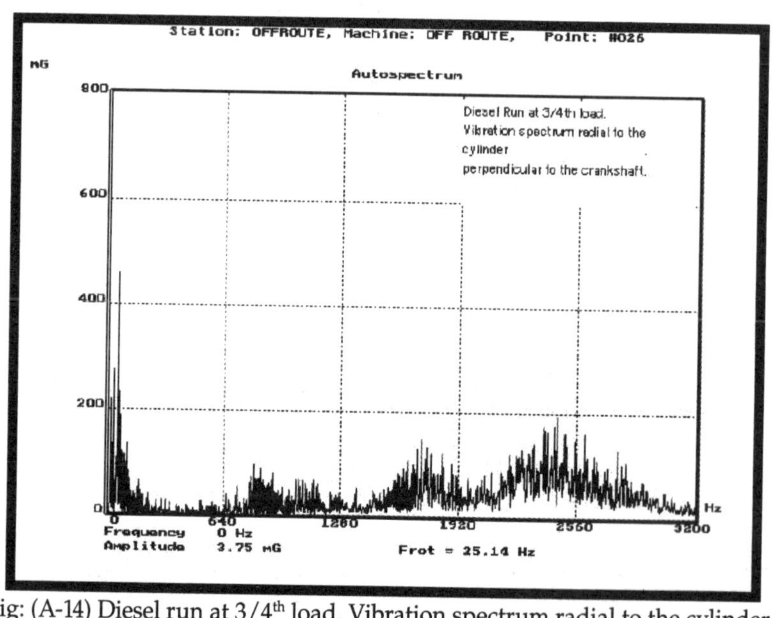

Fig: (A-14) Diesel run at 3/4th load. Vibration spectrum radial to the cylinder perpendicular to the crankshaft.

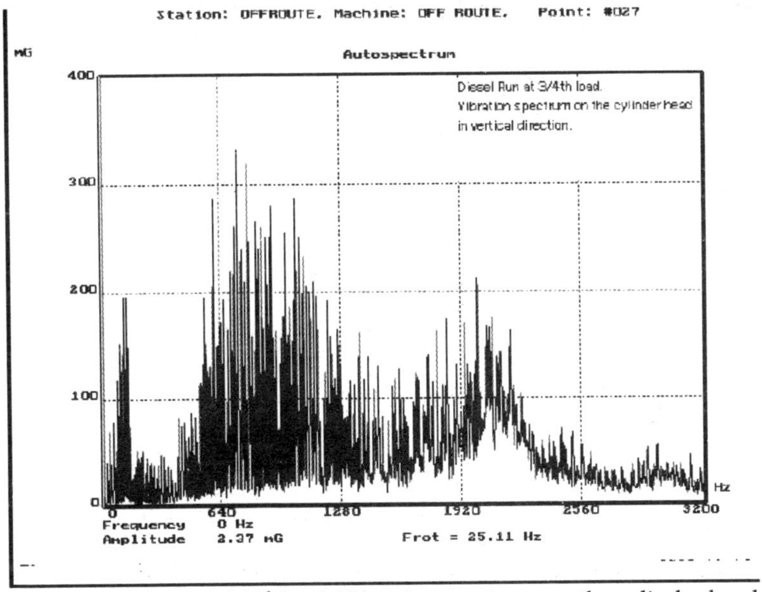

Fig:(A-15) Diesel run at 3/4th load. Vibration spectrum on the cylinder head in vertical direction.

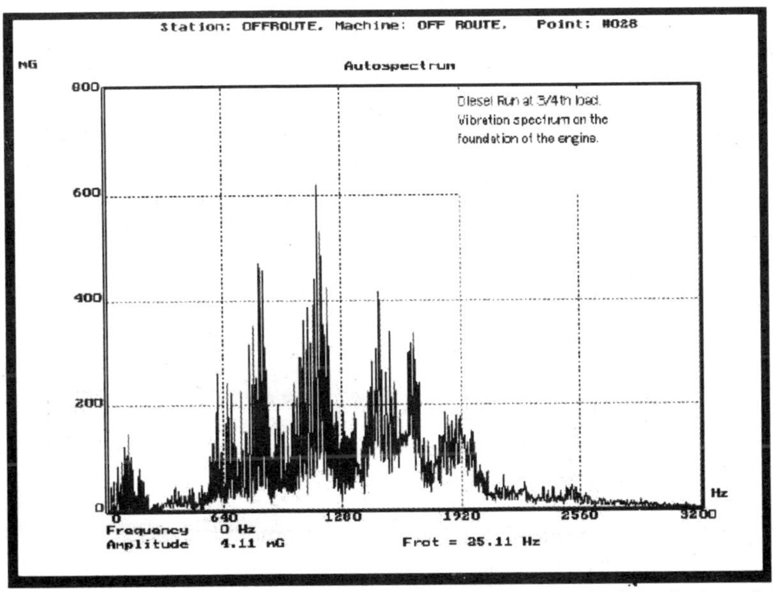

Fig: (A-16) Diesel run at 3/4th load. Vibration spectrum on the foundation of the engine.

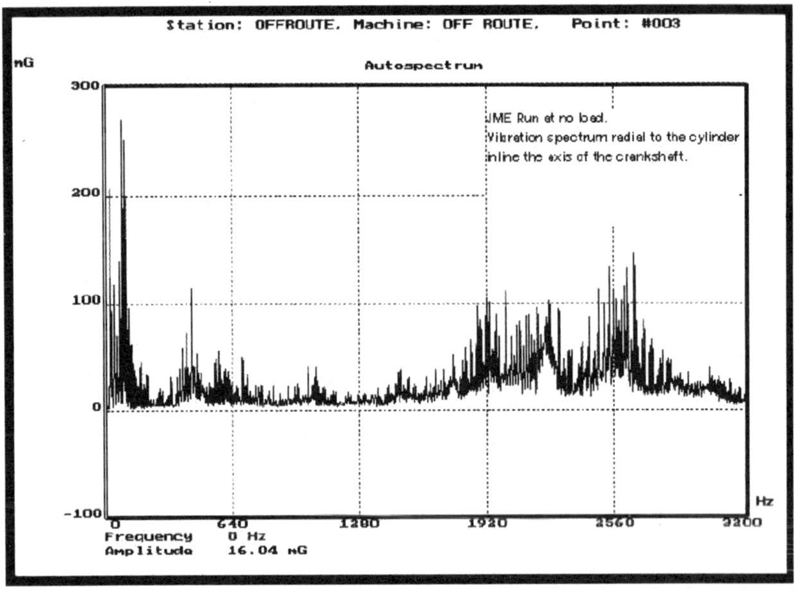

Fig:(A-17) JME run at no load. Vibration spectrum radial to the cylinder inline the axis of the crankshaft.

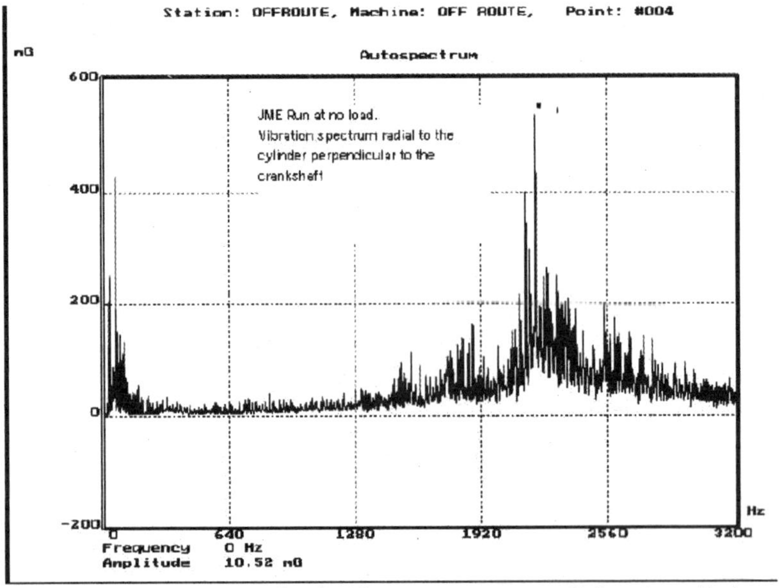

Fig: (A-18) JME run at no load. Vibration spectrum radial to the cylinder perpendicular to the crankshaft.

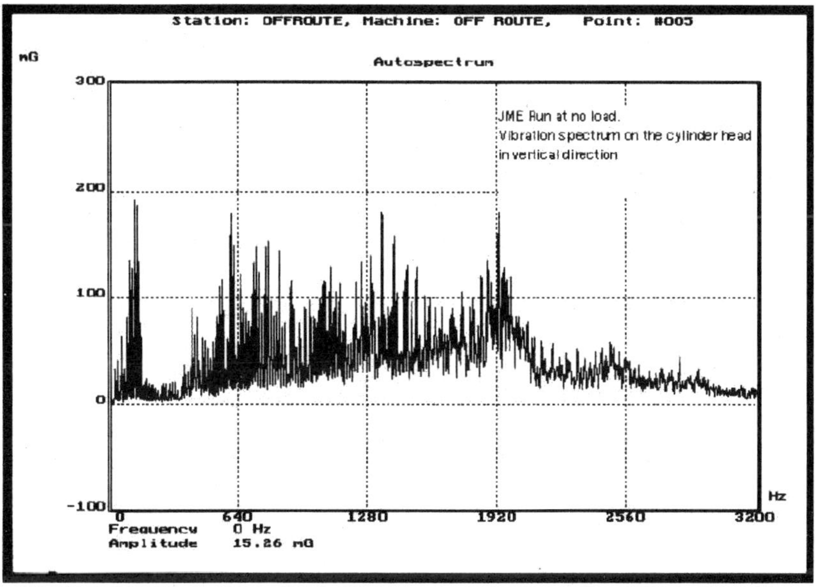

Fig:(A-19) JME run at no load. Vibration spectrum on the cylinder head in vertical direction.

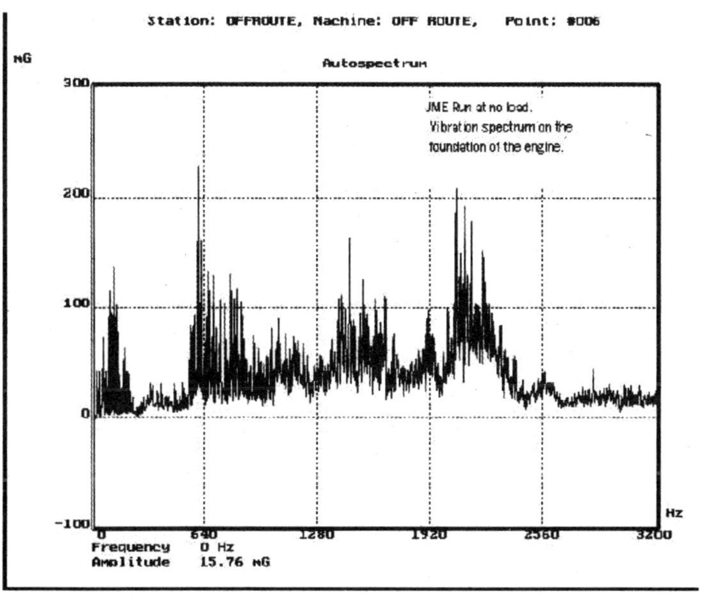

Fig: (A-20) JME run at no load. Vibration spectrum on the foundation of the engine.

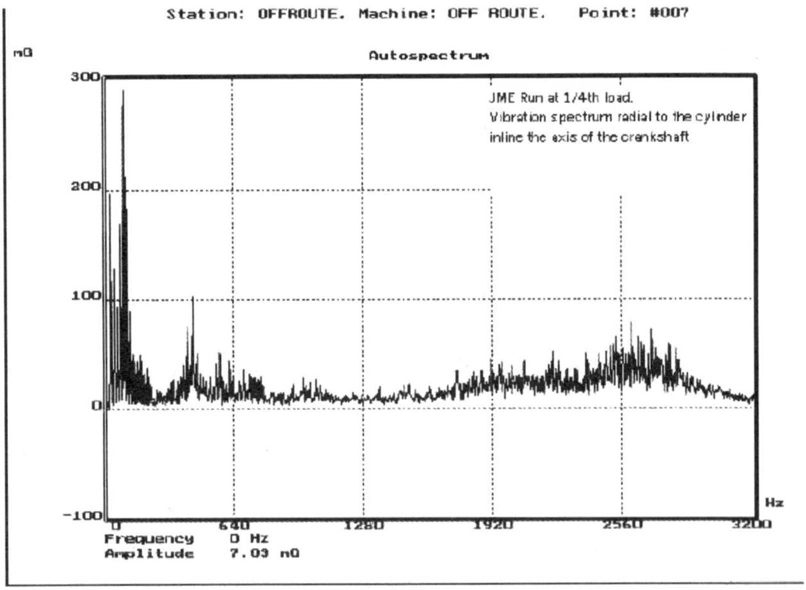

Fig:(A-21) JME run at 1/4th load. Vibration spectrum radial to the cylinder inline the axis of the crankshaft.

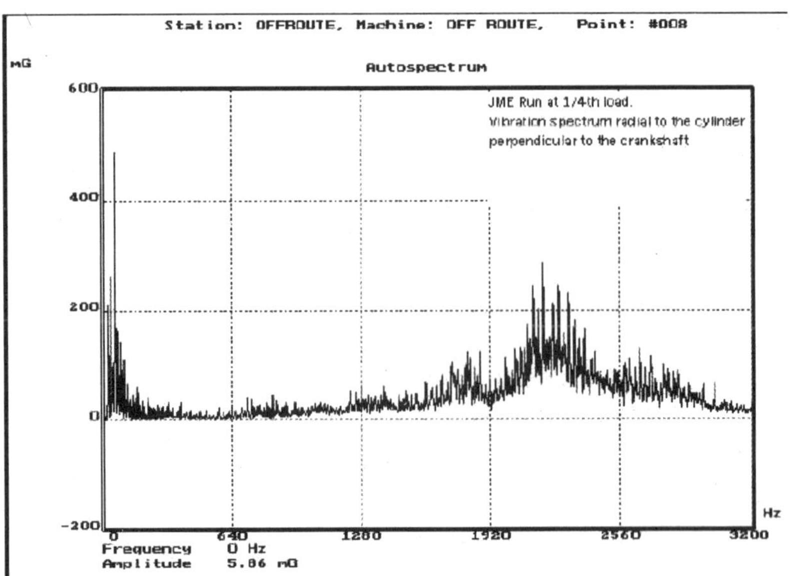

Fig: (A-22) JME run at 1/4th load. Vibration spectrum radial to the cylinder perpendicular to the crankshaft.

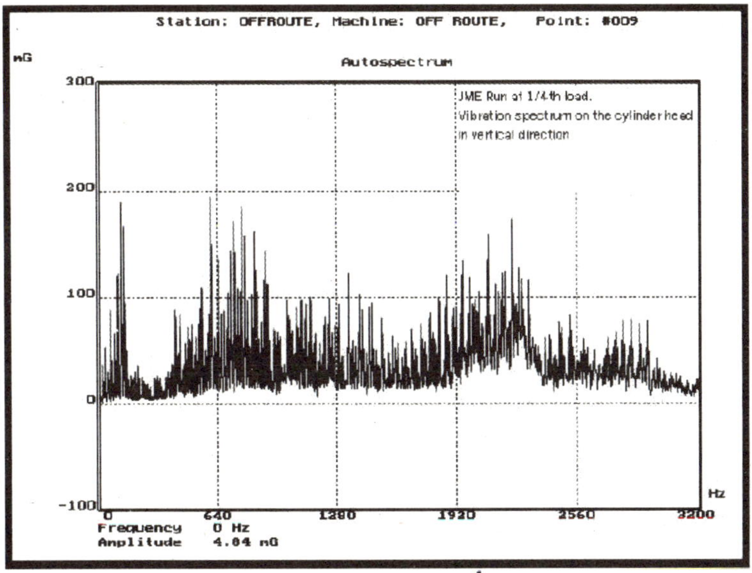

Fig: (A-23) JME run at 1/4th load. Vibration spectrum on the cylinder head in vertical direction.

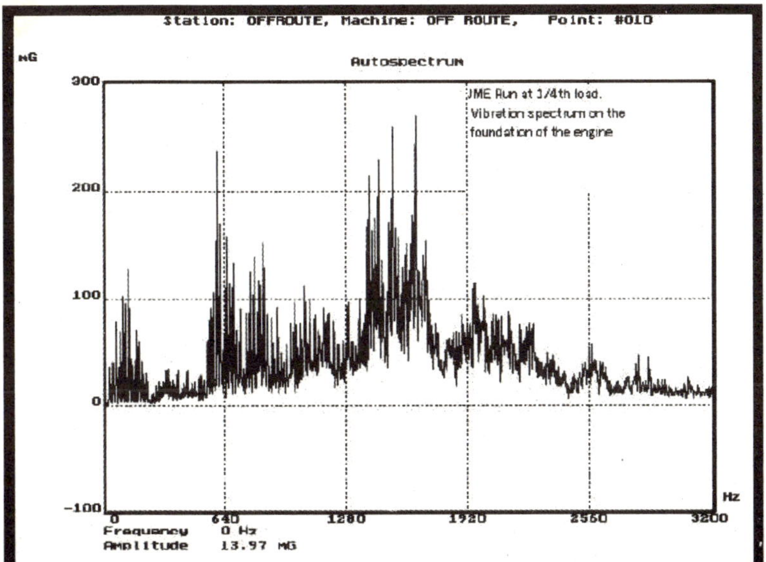

Fig:(A-24) JME run at 1/4th load. Vibration spectrum on the foundation of the engine.

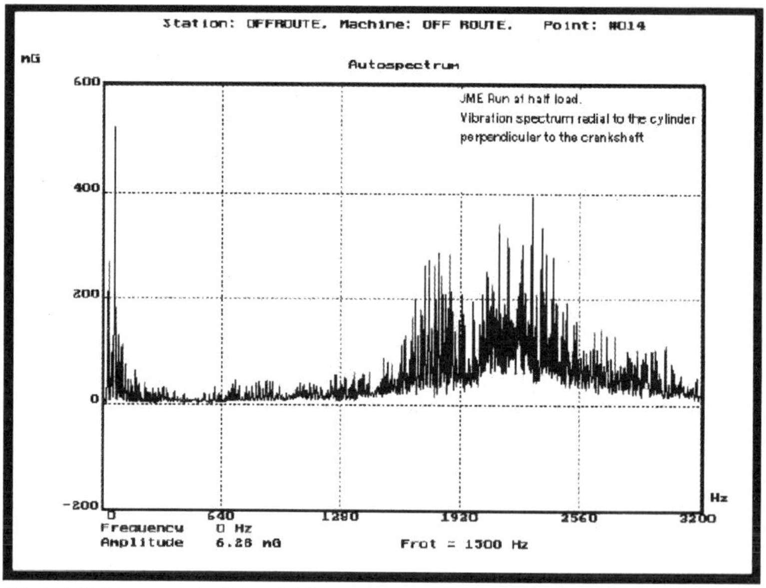

Fig:(A-25) JME run at half load. Vibration spectrum radial to the cylinder inline the axis of the crankshaft.

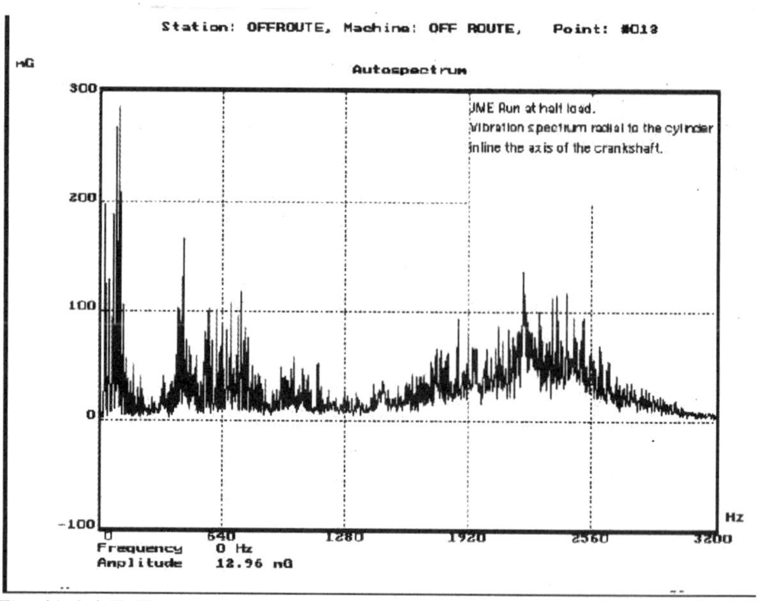

Fig: (A-26) JME run at half load. Vibration spectrum radial to the cylinder perpendicular to the crankshaft.

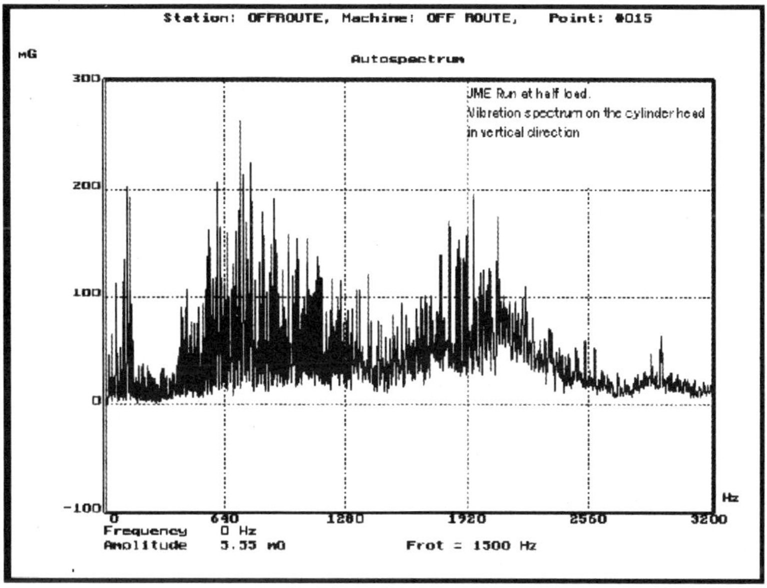

Fig:(A-27) JME run at half load. Vibration spectrum on the cylinder head in vertical direction.

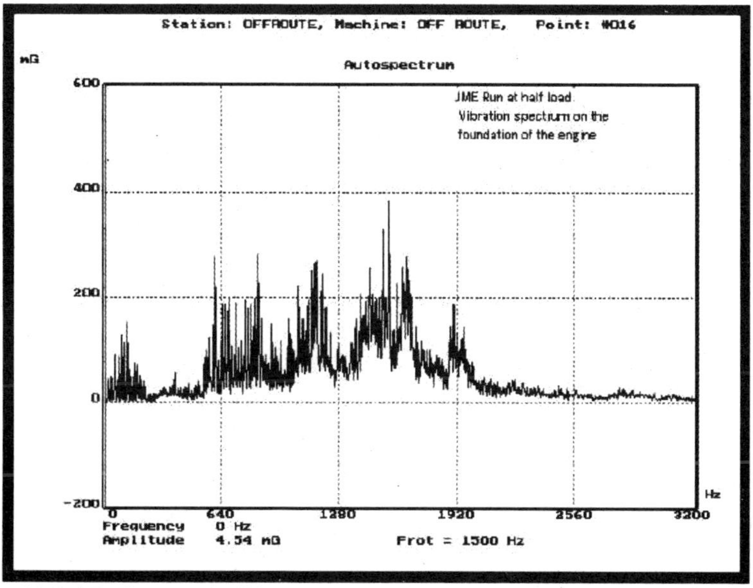

Fig: (A-28) JME run at half load. Vibration spectrum on the foundation of the engine.

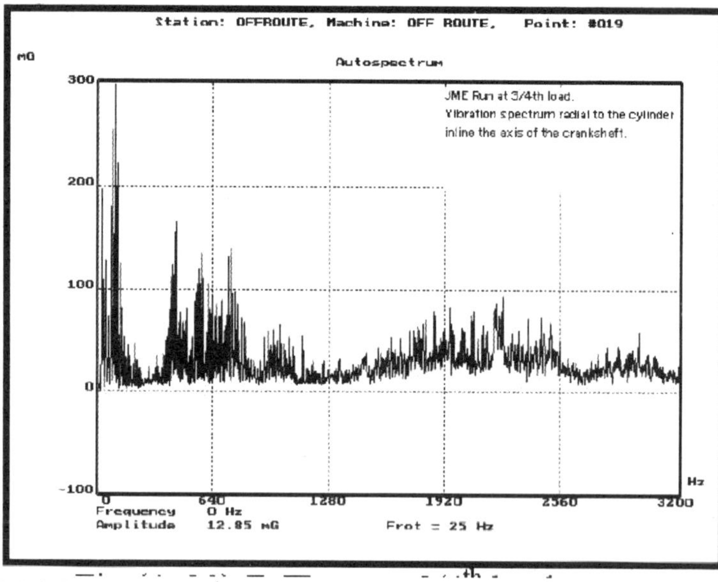

Fig: (A-29) JME run at 3/4th load. Vibration spectrum radial to the cylinder inline the axis of the crankshaft.

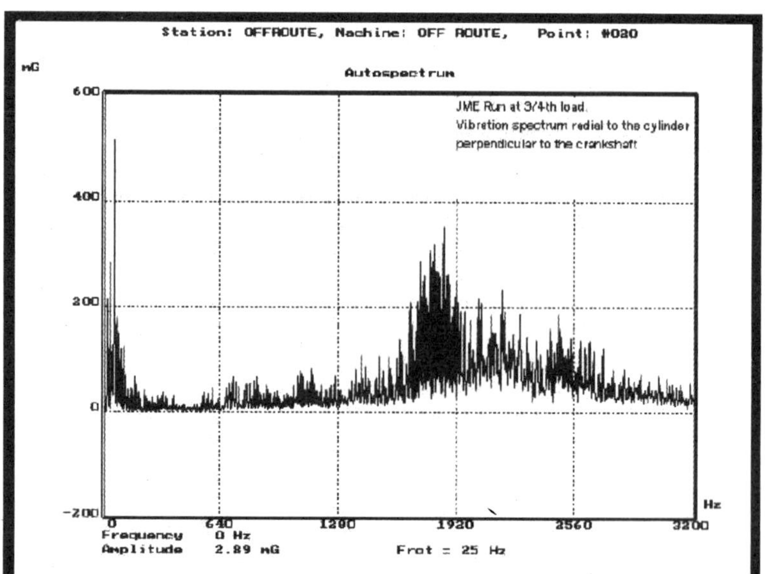

Fig:(A-30) JME run at 3/4th load. Vibration spectrum radial to the cylinder perpendicular to the crankshaft.

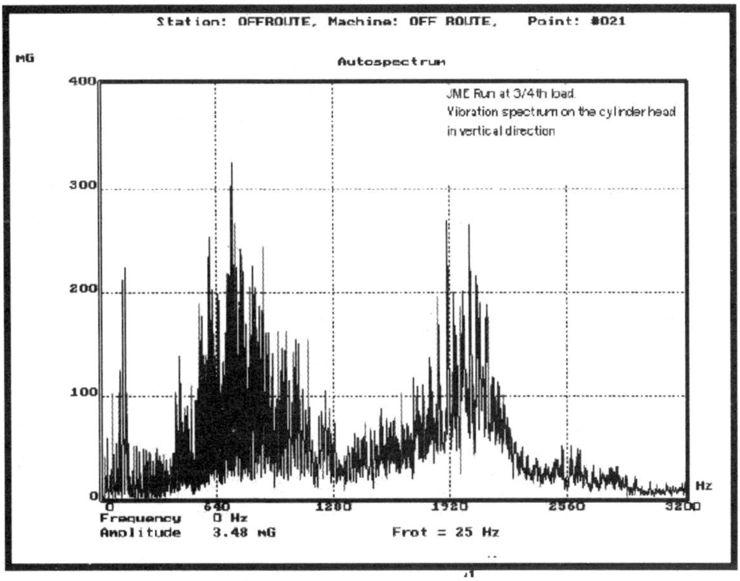

Fig: (A-31) JME run at 3/4th load. Vibration spectrum on the cylinder head in vertical direction.

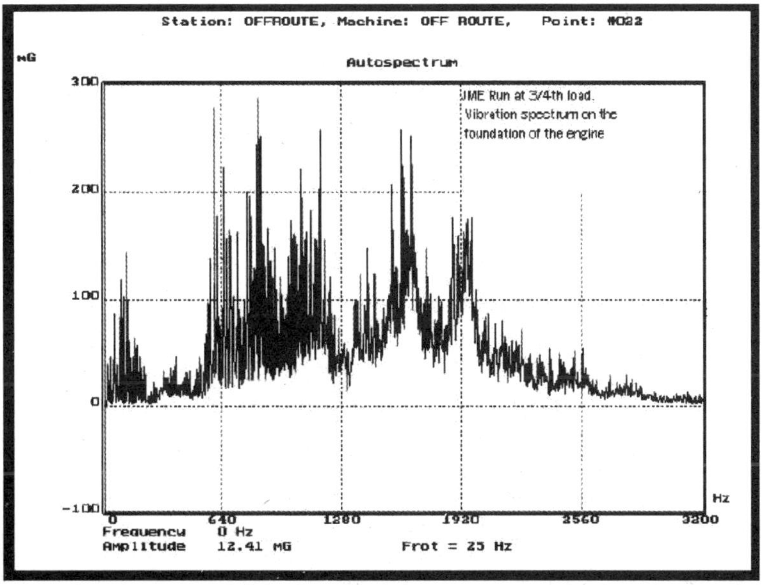

Fig:(A-32) JME run at 3/4th load. Vibration spectrum on the foundation of the engine.

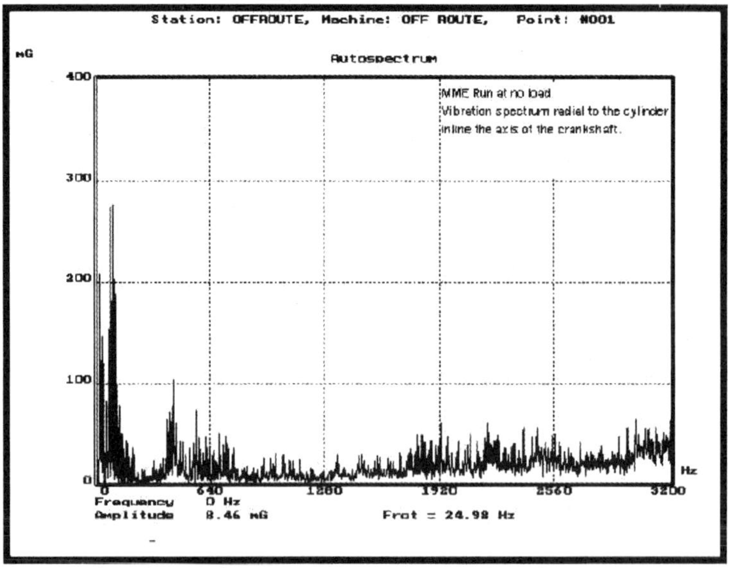

Fig: (A-33) MME run at no load. Vibration spectrum radial to the cylinder inline the axis of the crankshaft.

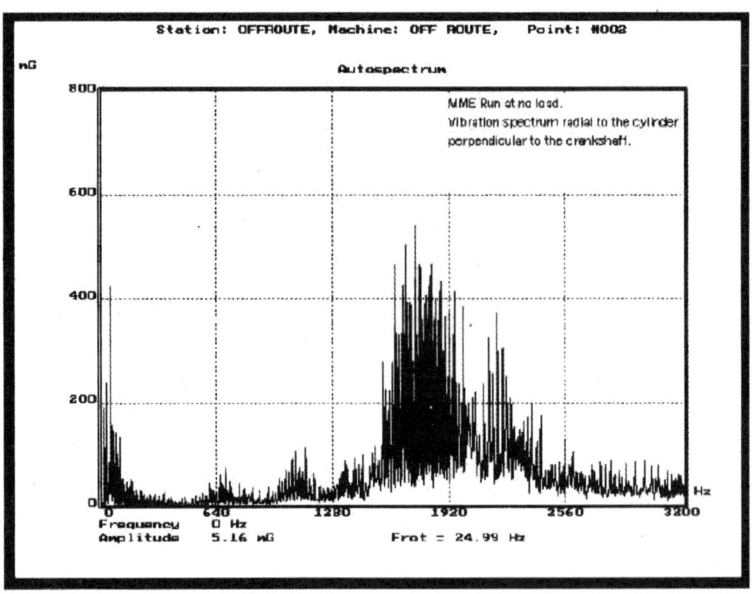

Fig:(A-34) MME run at no load. Vibration spectrum radial to the cylinder perpendicular to the crankshaft.

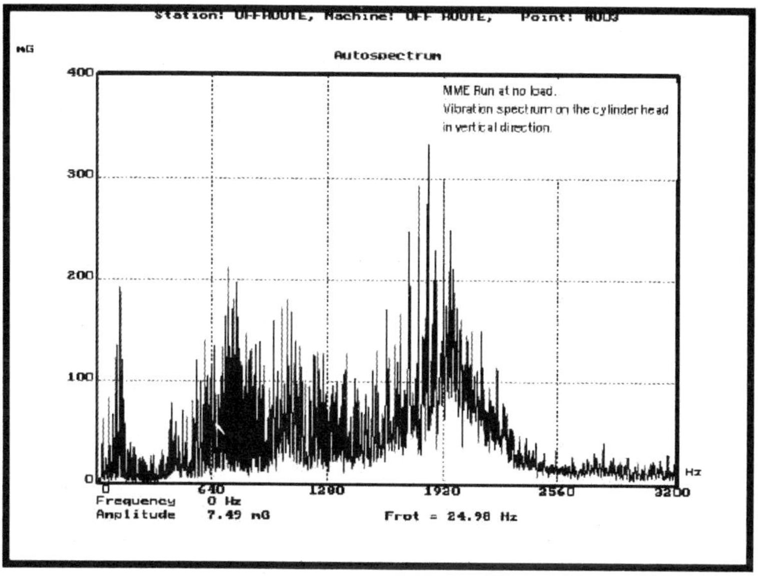

Fig:(A-35) MME run at no load. Vibration spectrum on the cylinder head in vertical direction.

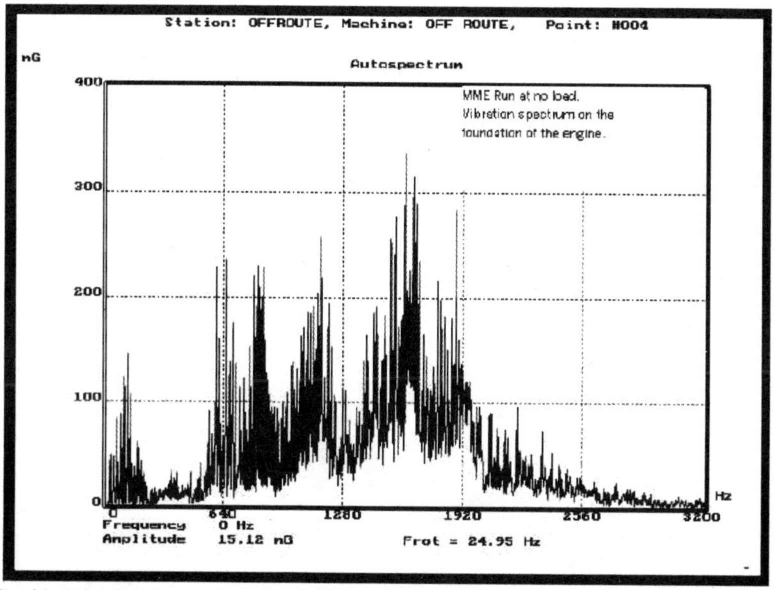

Fig:(A-36) MME run at no load. Vibration spectrum on the foundation of the engine

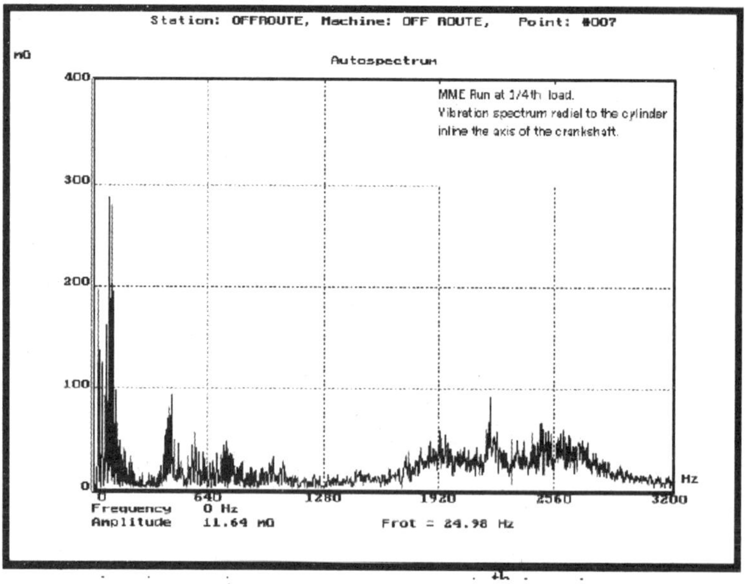

Fig:(A-37) MME run at 1/4th load. Vibration spectrum radial to the cylinder inline the axis of the crankshaft.

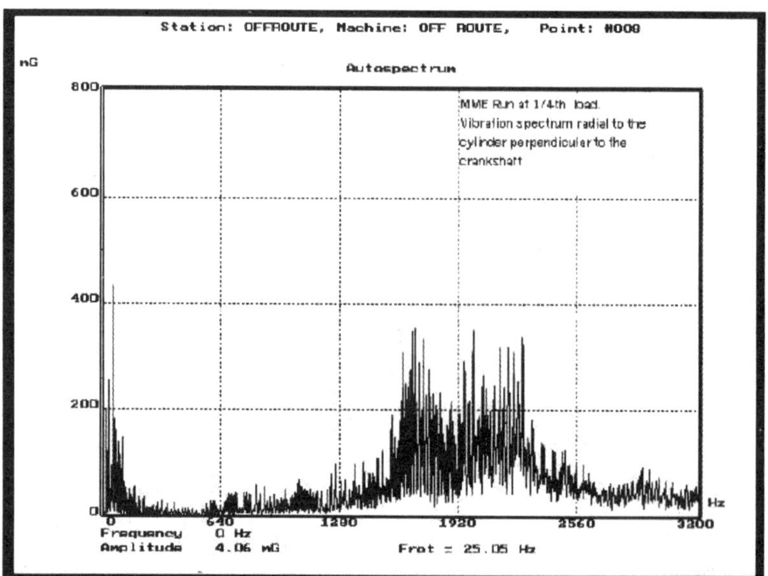

Fig:(A-38) MME run at 1/4th load. Vibration spectrum radial to the cylinder perpendicular to the crankshaft.

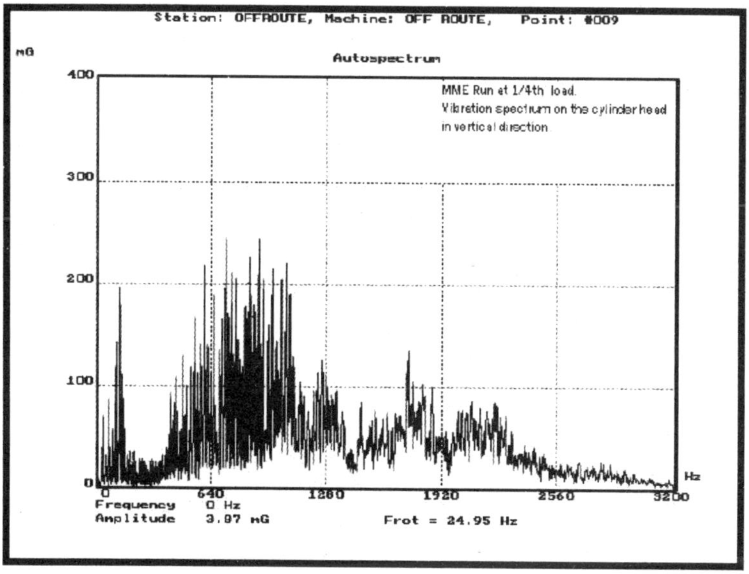

Fig:(A-39) MME run at 1/4th load. Vibration spectrum on the cylinder head in vertical direction.

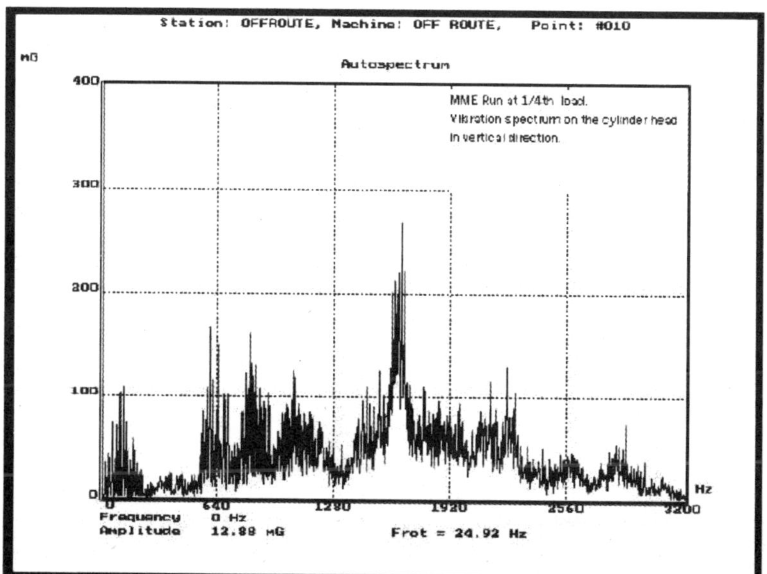

Fig:(A-40) MME run at 1/4th load. Vibration spectrum on the foundation of the engine.

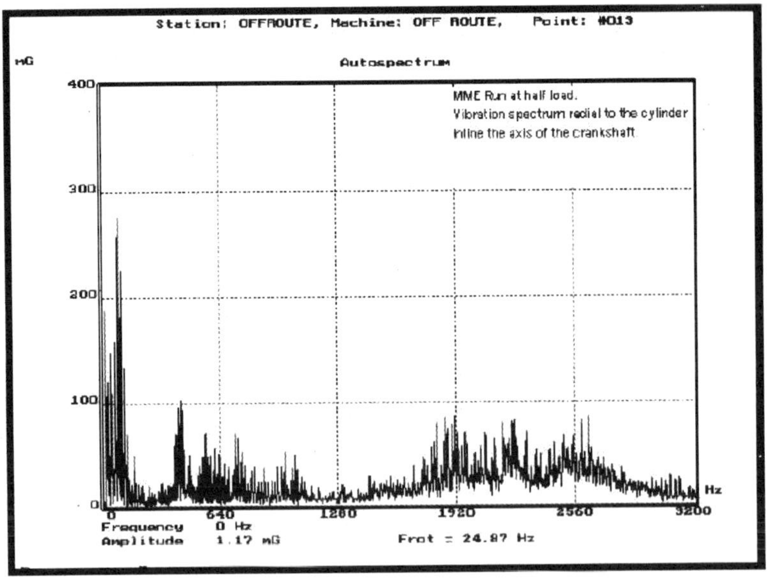

Fig:(A-41) MME run at half load. Vibration spectrum radial to the cylinder inline the axis of the crankshaft.

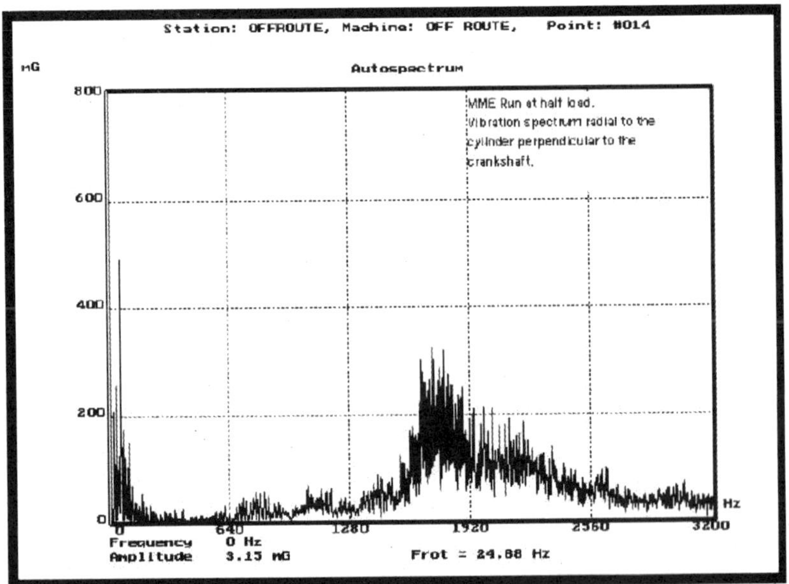

Fig:(A-42) MME run at half load. Vibration spectrum radial to the cylinder perpendicular to the crankshaft.

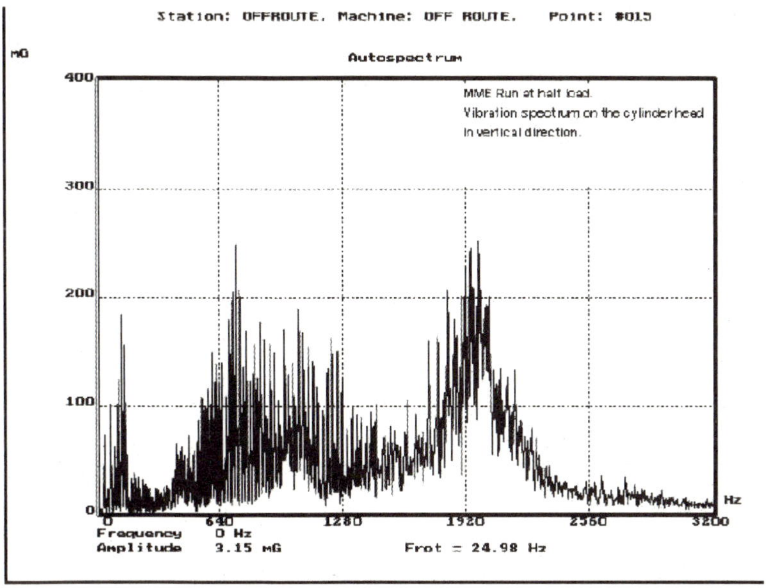

Fig:(A-43) MME run at half load. Vibration spectrum on the cylinder head in vertical direction.

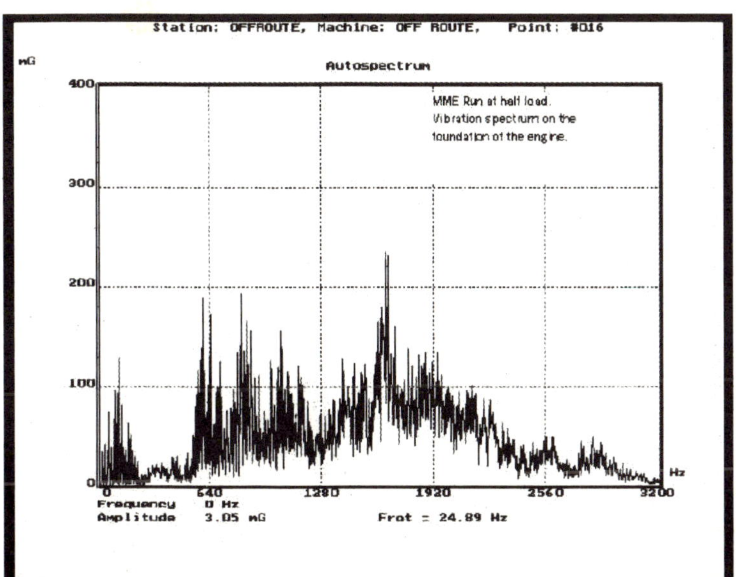

Fig:(A-44) MME run at half load. Vibration spectrum on the foundation of the engine.

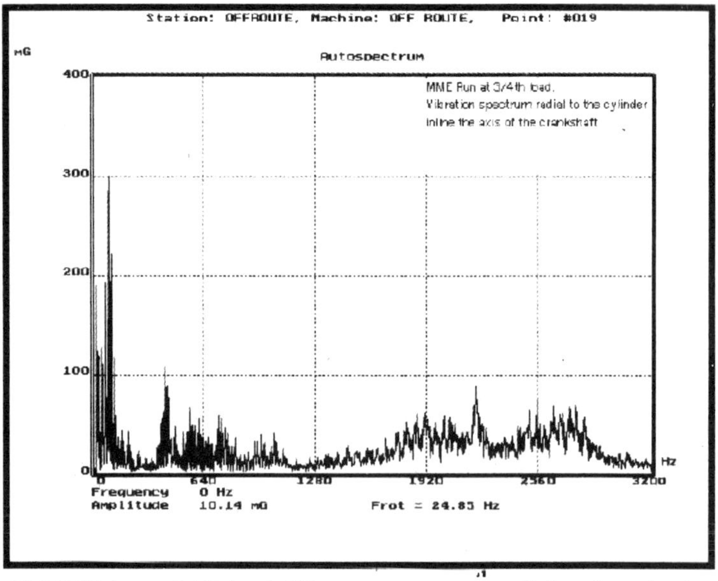

Fig:(A-45) MME run at 3/4th load. Vibration spectrum radial to the cylinder inline the axis of the crankshaft.

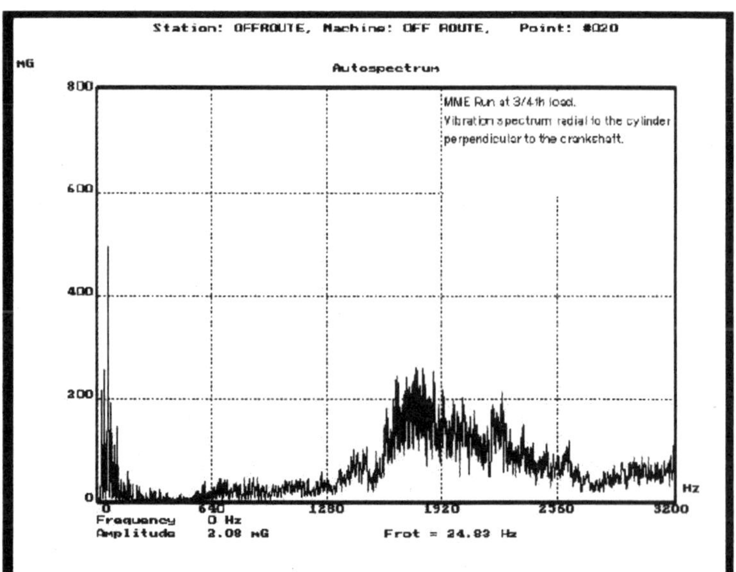

Fig:(A-46) MME run at 3/4th load. Vibration spectrum radial to the cylinder perpendicular to the crankshaft.

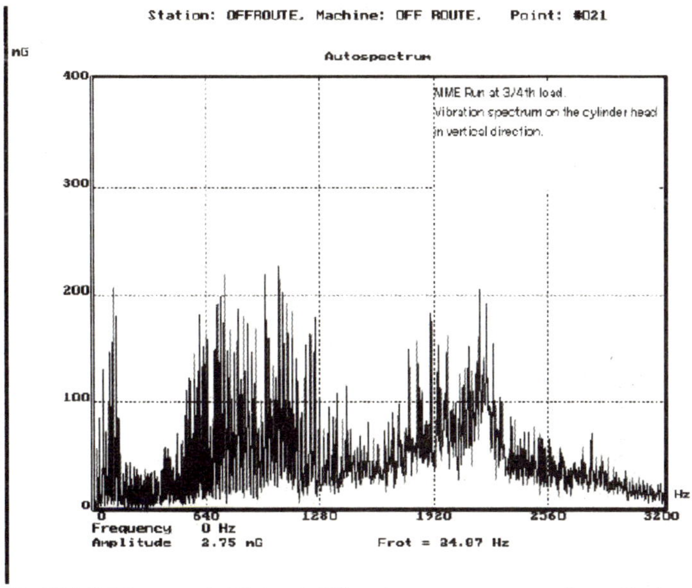

Fig:(A-47) MME run at 3/4th load. Vibration spectrum on the cylinder head in vertical direction.

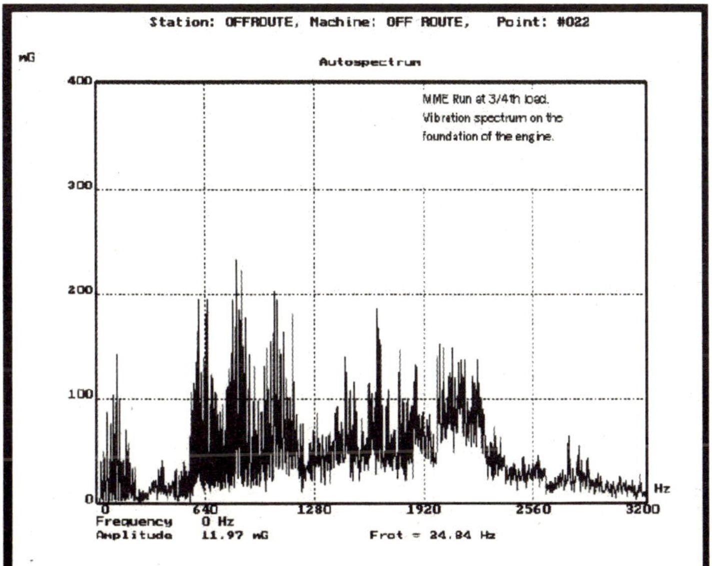

Fig:(A-48) MME run at 3/4th load. Vibration spectrum on the foundation of the engine.

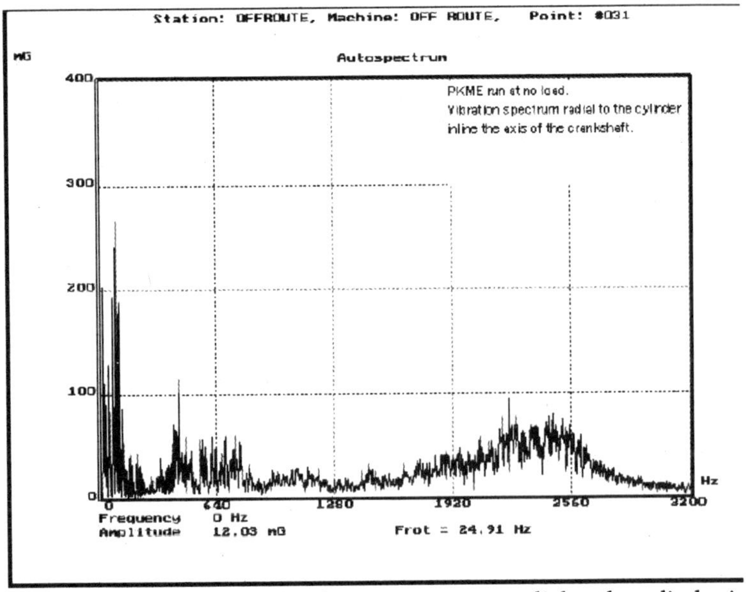

Fig:(A-49) PKME run at no load. Vibration spectrum radial to the cylinder inline the axis of the crankshaft.

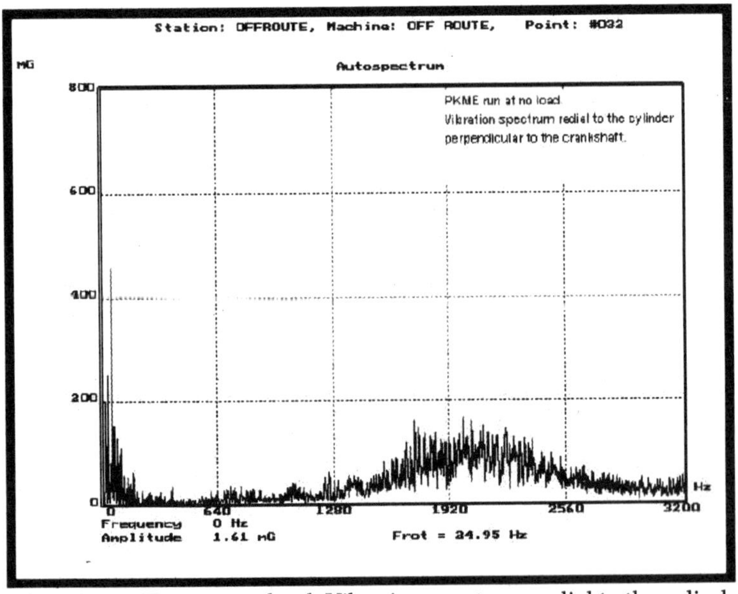

Fig:(A-50) PKME run at no load. Vibration spectrum radial to the cylinder perpendicular to the crankshaft.

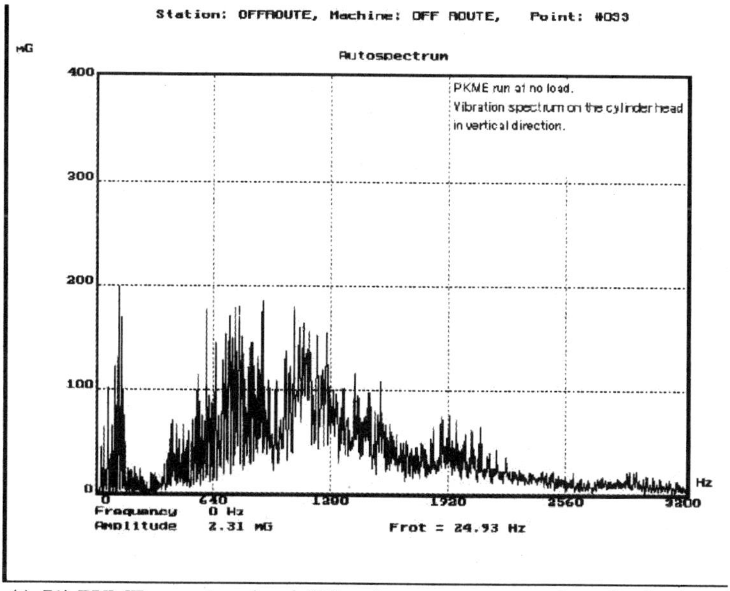

Fig:(A-51) PKME run at no load. Vibration spectrum on the cylinder head in vertical direction.

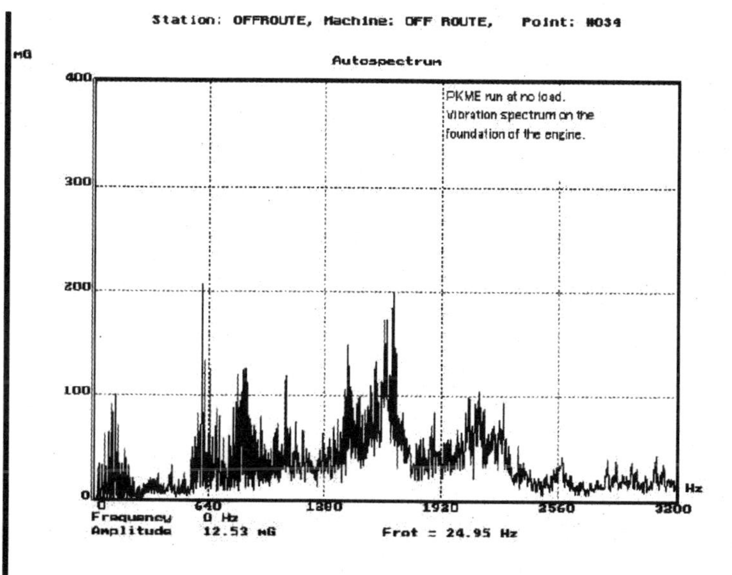

Fig:(A-52) PKME run at no load. Vibration spectrum on the foundation of the engine.

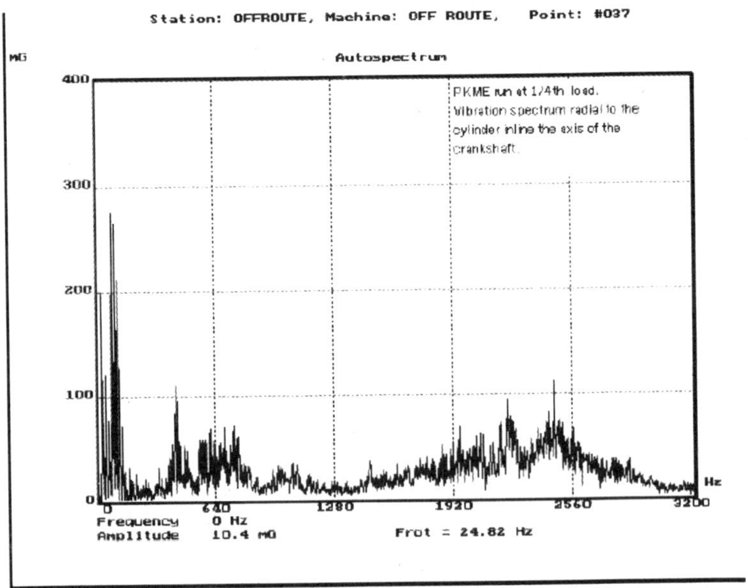

Fig:(A-53) PKME run at 1/4th load. Vibration spectrum radial to the cylinder inline the axis of the crankshaft.

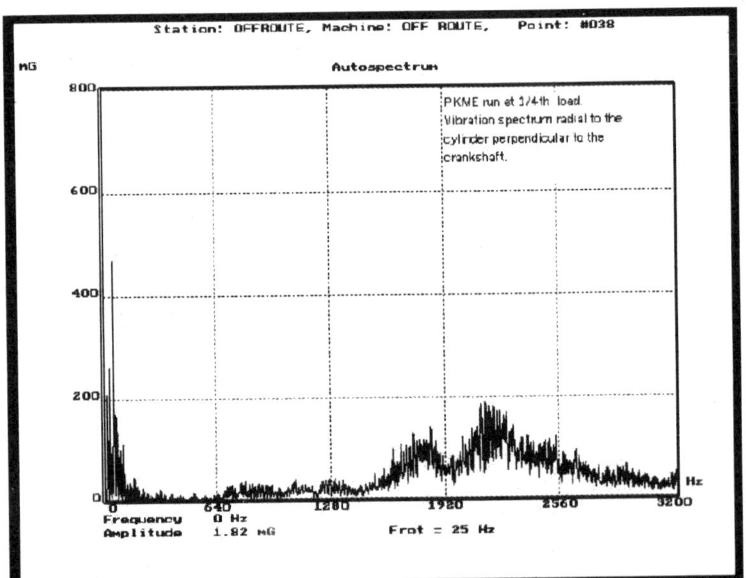

Fig:(A-54) PKME run at 1/4th load. Vibration spectrum radial to the cylinder perpendicular to the crankshaft.

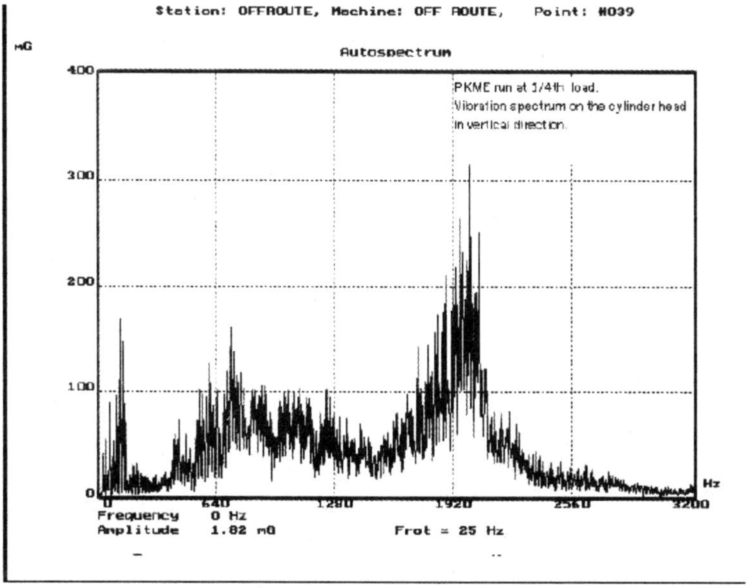

Fig:(A-55) PKME run at $1/4^{th}$ load. Vibration spectrum on the cylinder head in vertical direction.

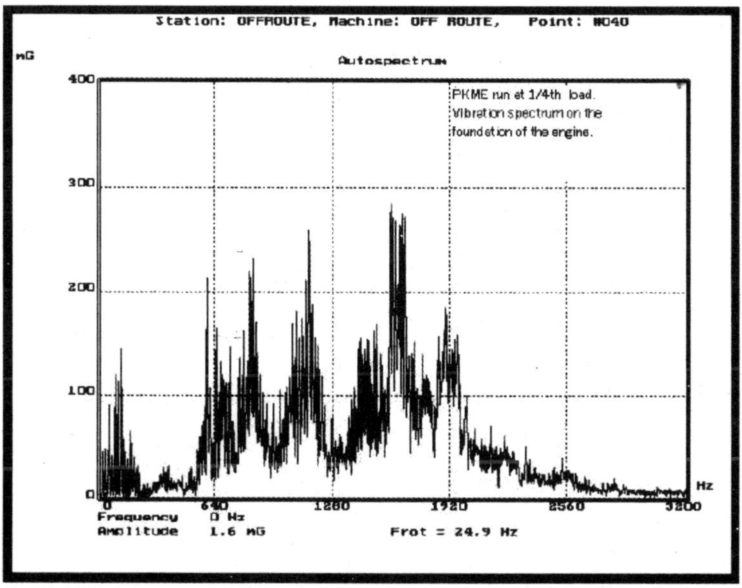

Fig:(A-56) PKME run at $1/4^{th}$ load. Vibration spectrum on the foundation of the engine.

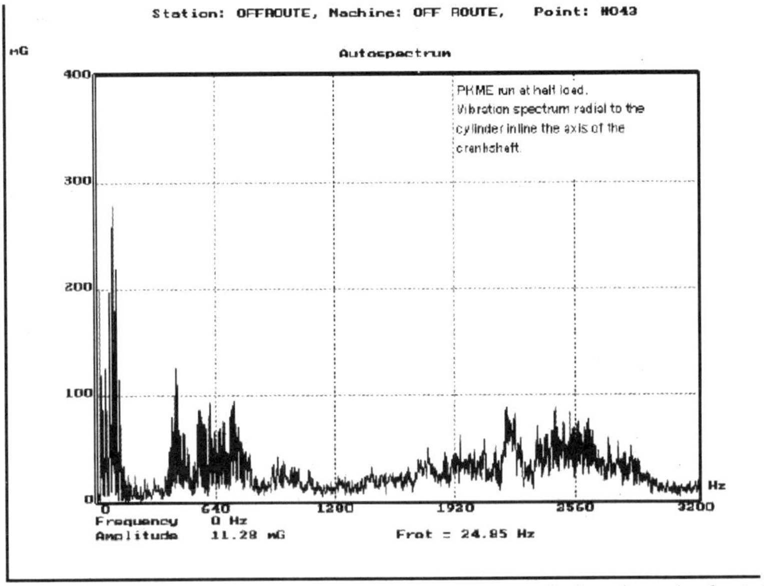

Fig:(A-57) PKME run at half load. Vibration spectrum radial to the cylinder inline the axis of the crankshaft.

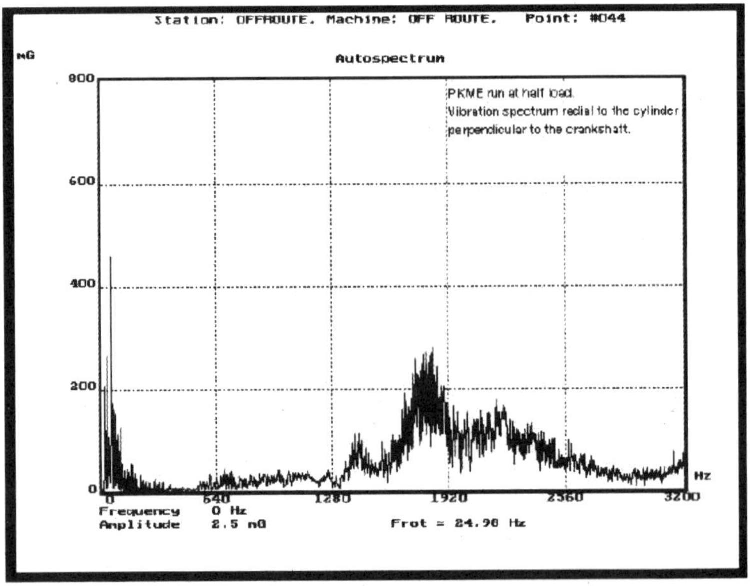

Fig:(A-58) PKME run at half load. Vibration spectrum radial to the cylinder perpendicular to the crankshaft.

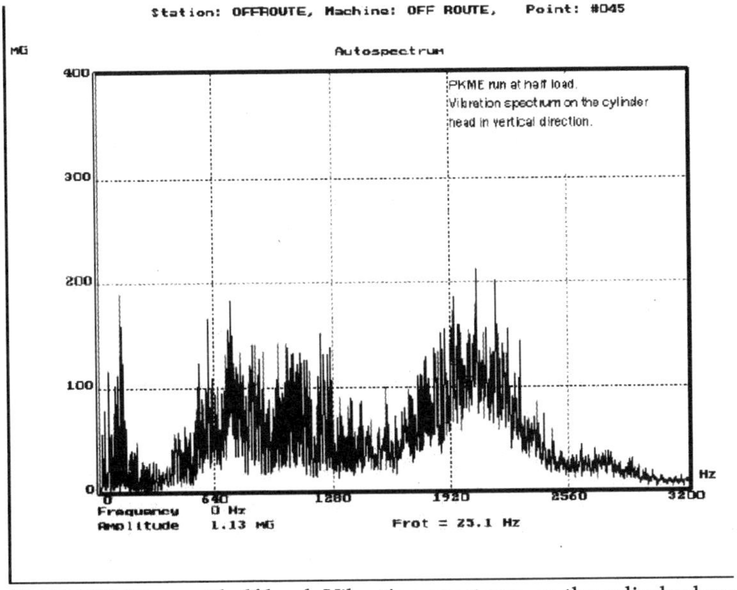

Fig:(A-59) PKME run at half load. Vibration spectrum on the cylinder head in vertical direction.

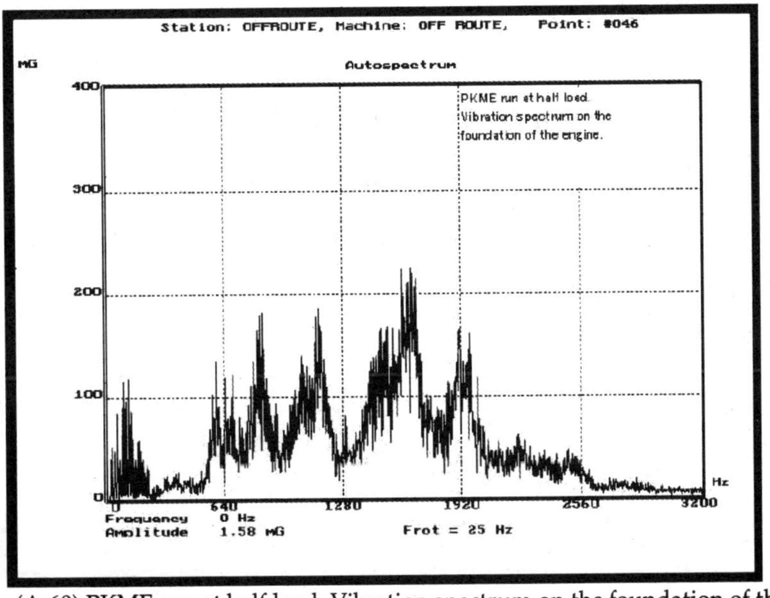

Fig:(A-60) PKME run at half load. Vibration spectrum on the foundation of the engine.

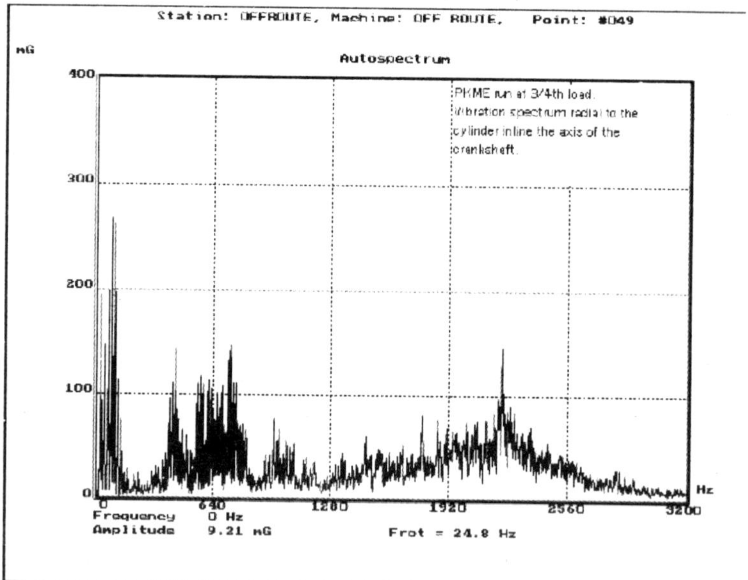

Fig:(A-61) PKME run at 3/4th load. Vibration spectrum radial to the cylinder inline the axis of the crankshaft.

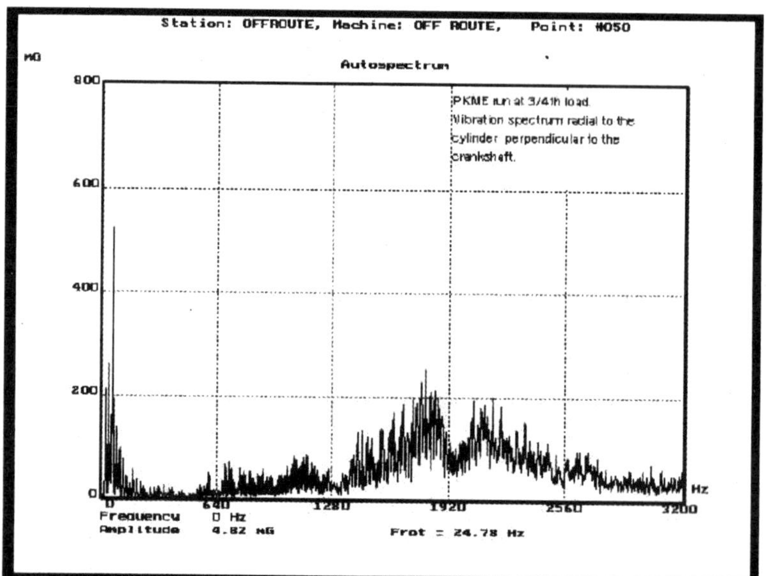

Fig:(A-62) PKME run at 3/4th load. Vibration spectrum radial to the cylinder perpendicular to the crankshaft.

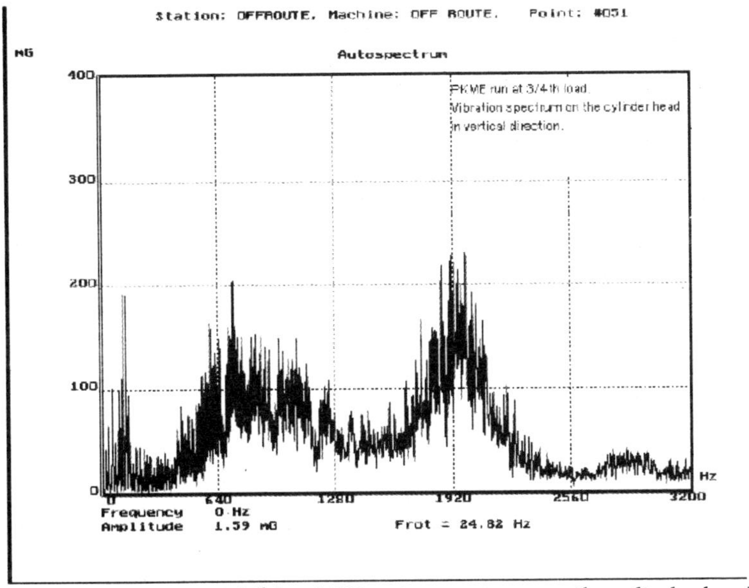

Fig:(A-63) PKME run at 3/4th load. Vibration spectrum on the cylinder head in vertical direction.

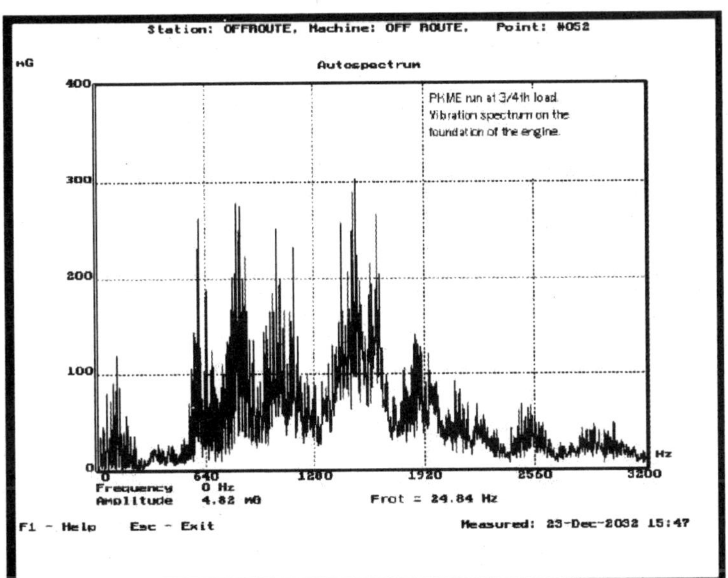

Fig:(A-64) PKME run at 3/4th load. Vibration spectrum on the foundation of the engine.

Appendix B

The time waveforms measured at various points on the engine at various loads have been appended below.

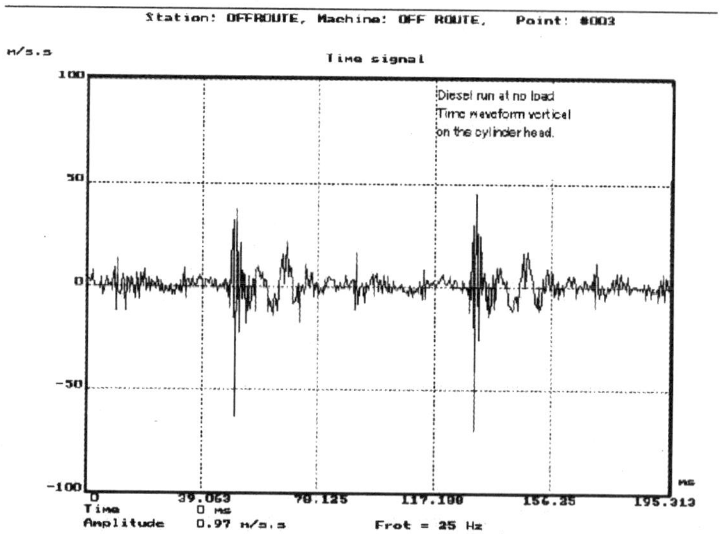

Fig:(B-1) Diesel run at no load. Time waveform vertical on the Cylinder head.

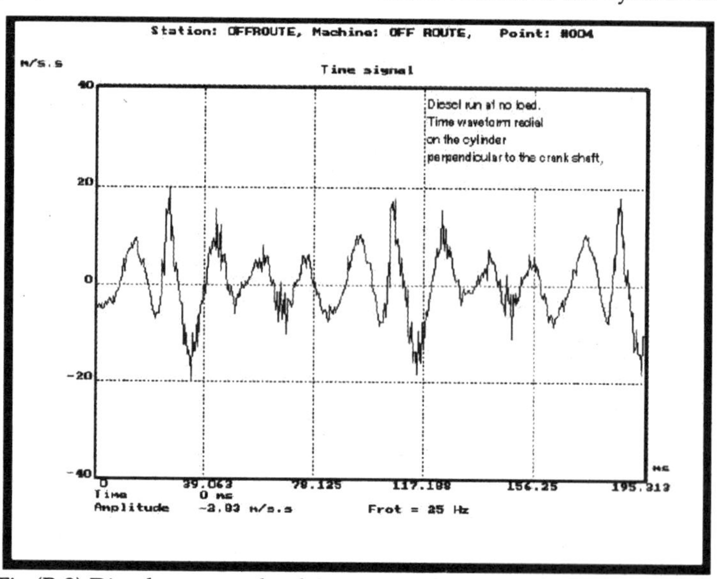

Fig:(B-2) Diesel run at no load. Time waveform radial on the cylinder perpendicular to the crankshaft.

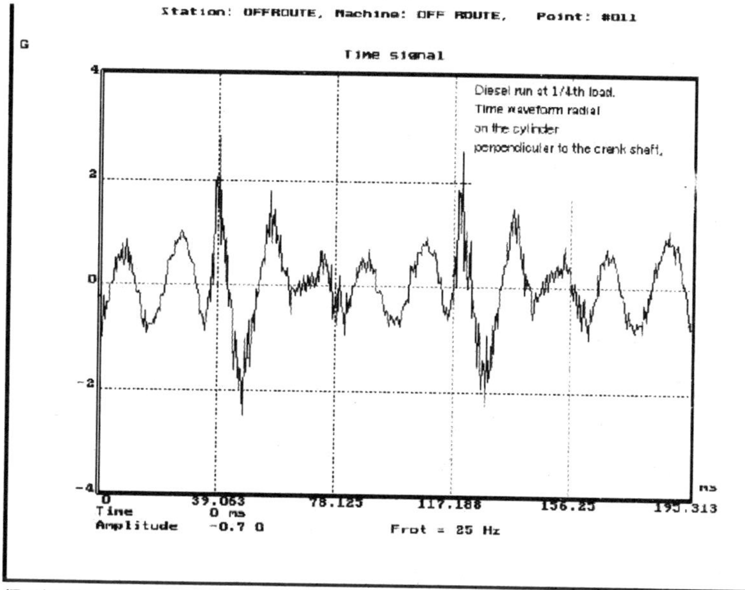

Fig:(B-3) Diesel run at 1/4th load. Time waveform vertical on the Cylinder head.

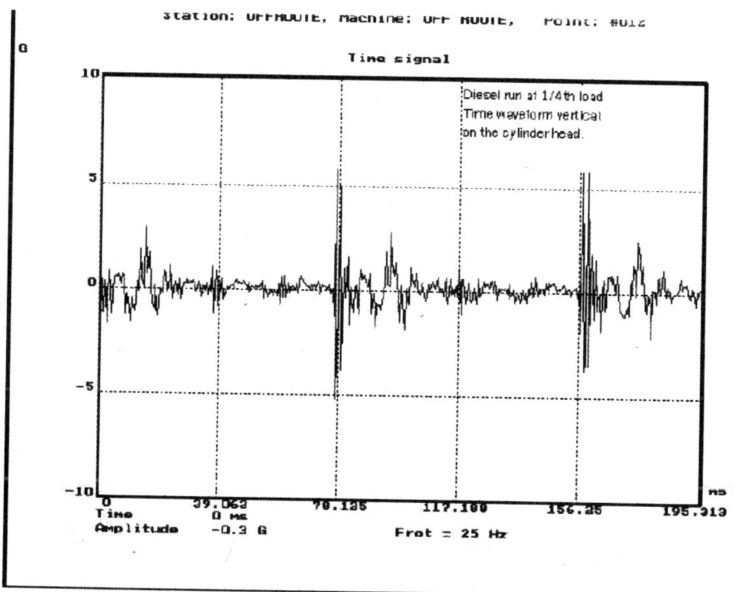

Fig:(B-4) Diesel run at 1/4th load. Time waveform radial on the cylinder perpendicular to the crankshaft.

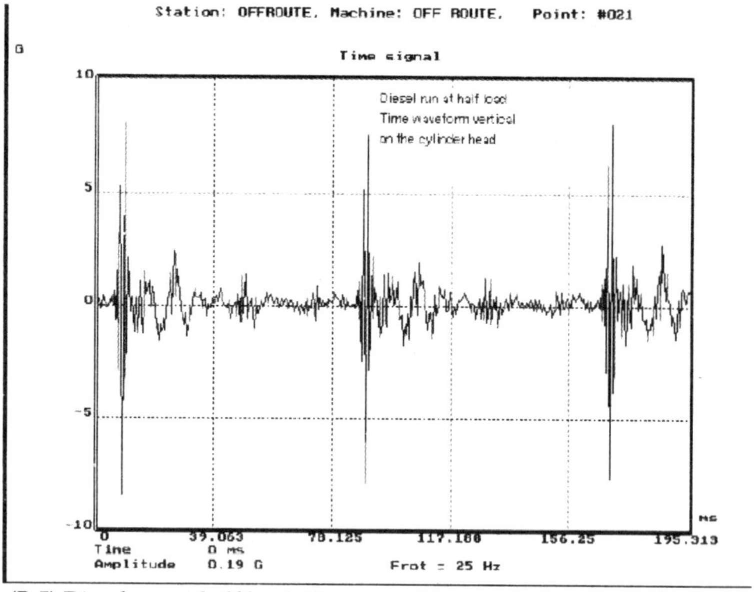

Fig:(B-5) Diesel run at half load. Time waveform vertical on the Cylinder head.

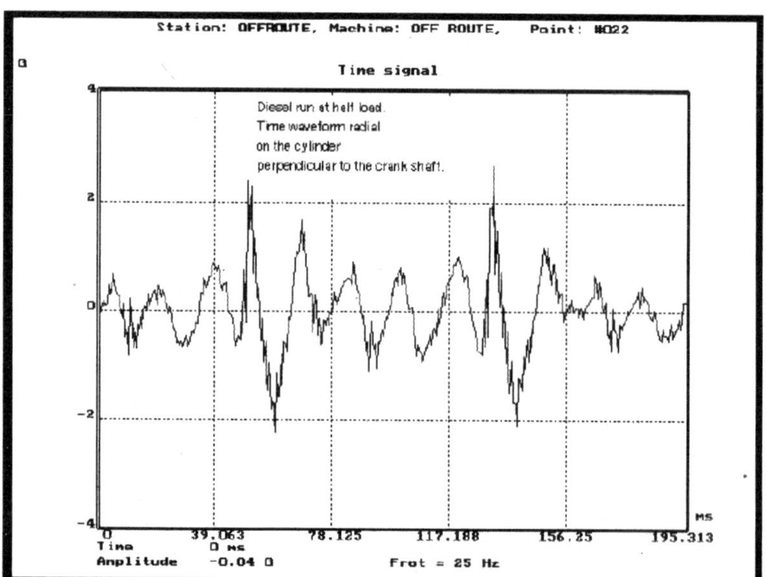

Fig:(B-6) Diesel run at half load. Time waveform radial on the cylinder perpendicular to the crankshaft.

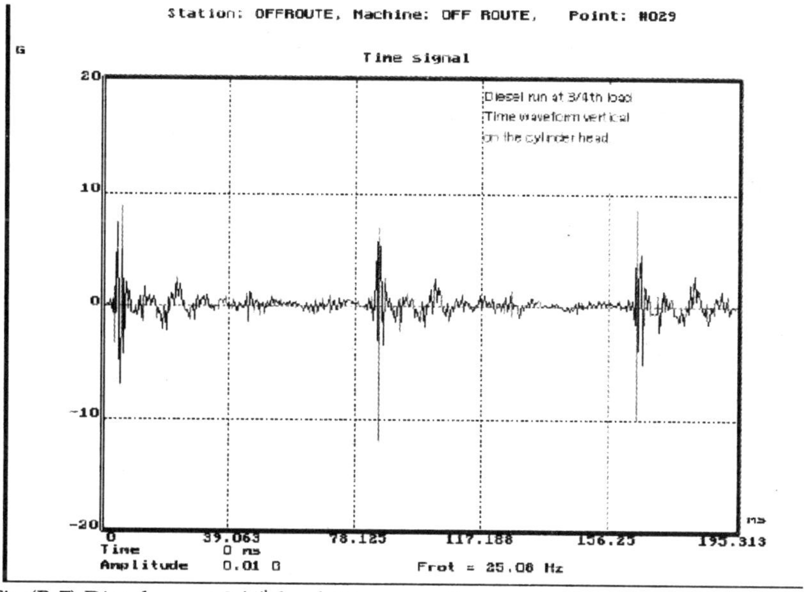

Fig:(B-7) Diesel run at 3/4th load. Time waveform vertical on the Cylinder head.

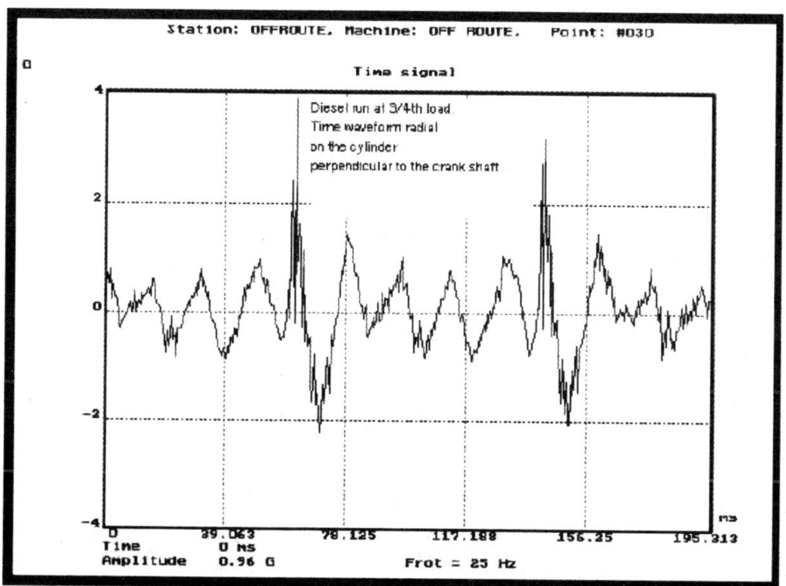

Fig:(B-8) Diesel run at 3/4th load. Time waveform radial on the cylinder perpendicular to the crankshaft.

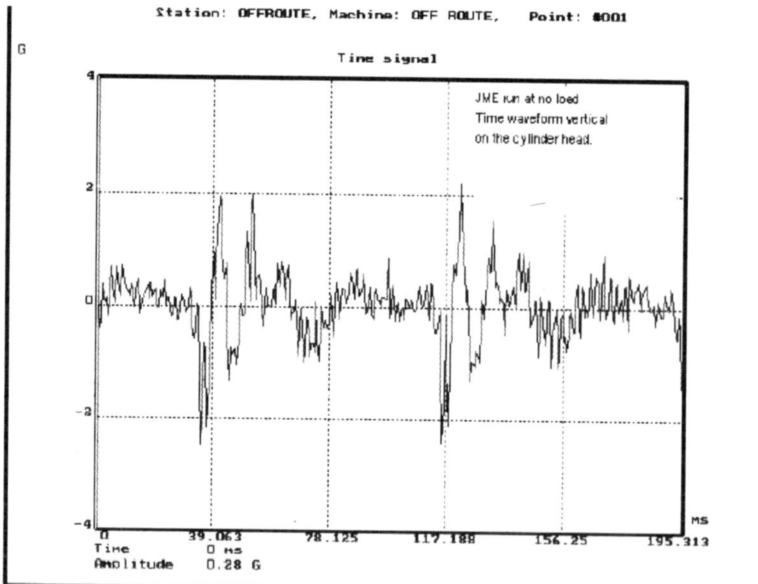

Fig:(B-9) JME run at no load. Time waveform vertical on the Cylinder head.

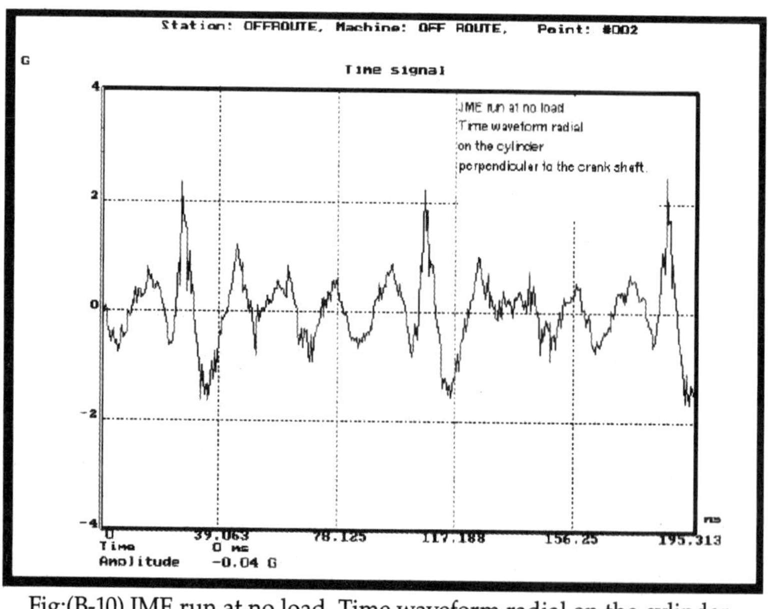

Fig:(B-10) JME run at no load. Time waveform radial on the cylinder perpendicular to the crankshaft.

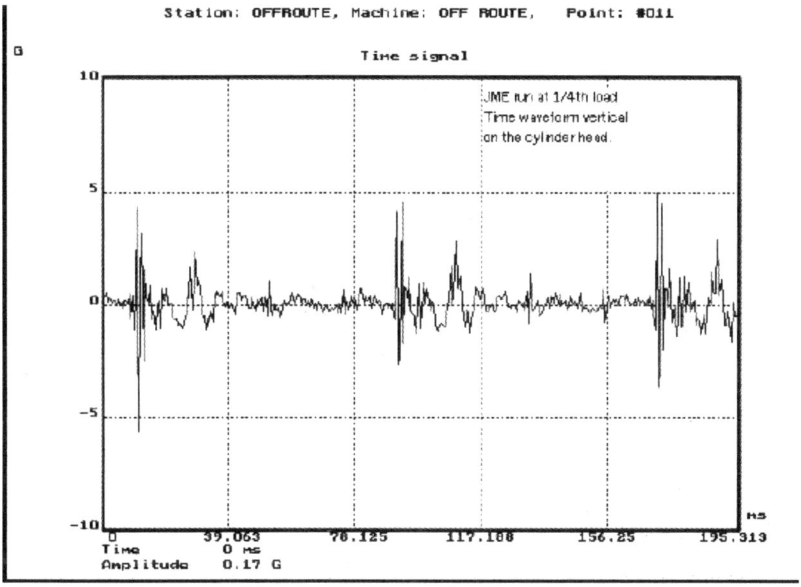

Fig:(B-11) JME run at 1/4th load. Time waveform vertical on the Cylinder head.

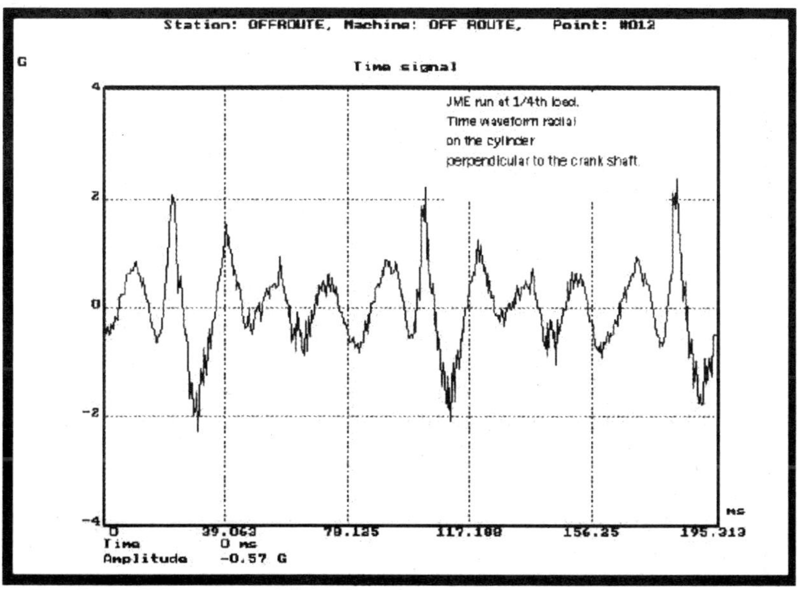

Fig:(B-12) JME run at 1/4th load. Time waveform radial on the cylinder perpendicular to the crankshaft.

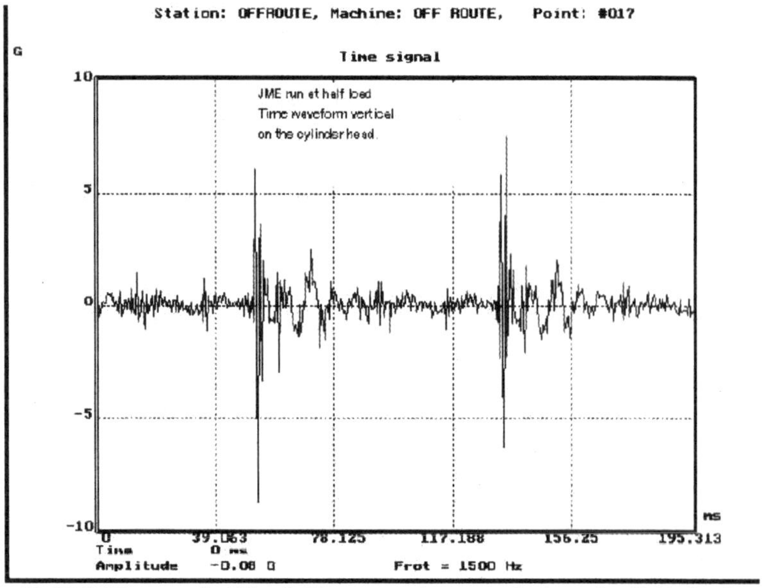

Fig:(B-13) JME run at half load. Time waveform vertical on the Cylinder head.

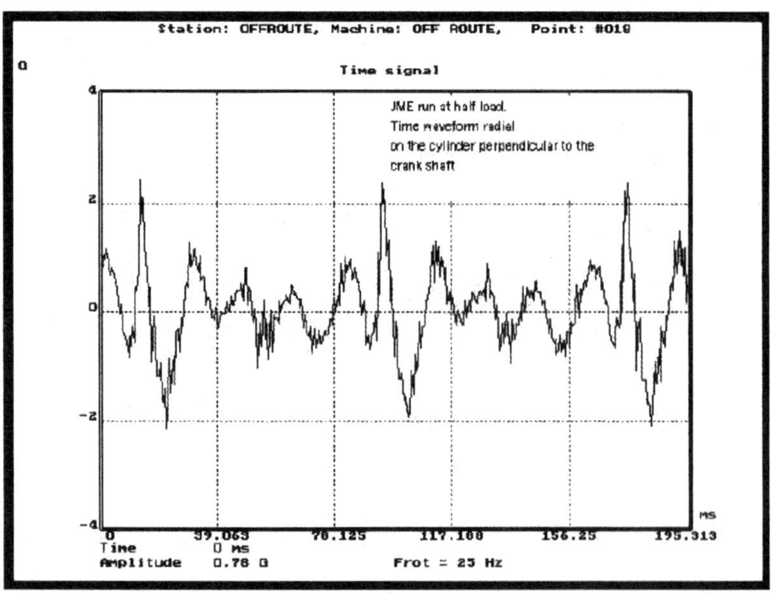

Fig:(B-14) JME run at half load. Time waveform radial on the cylinder perpendicular to the crankshaft.

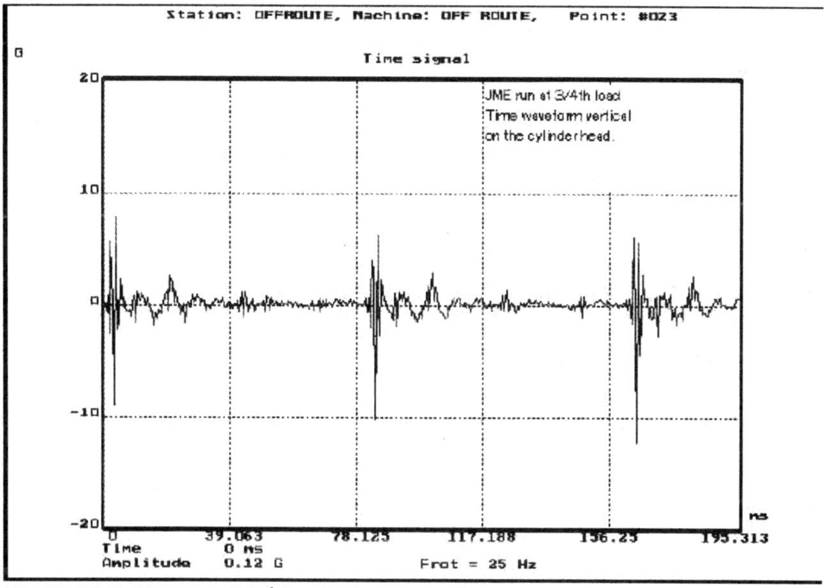

Fig:(B-15) JME run at 3/4th load. Time waveform vertical on the Cylinder head.

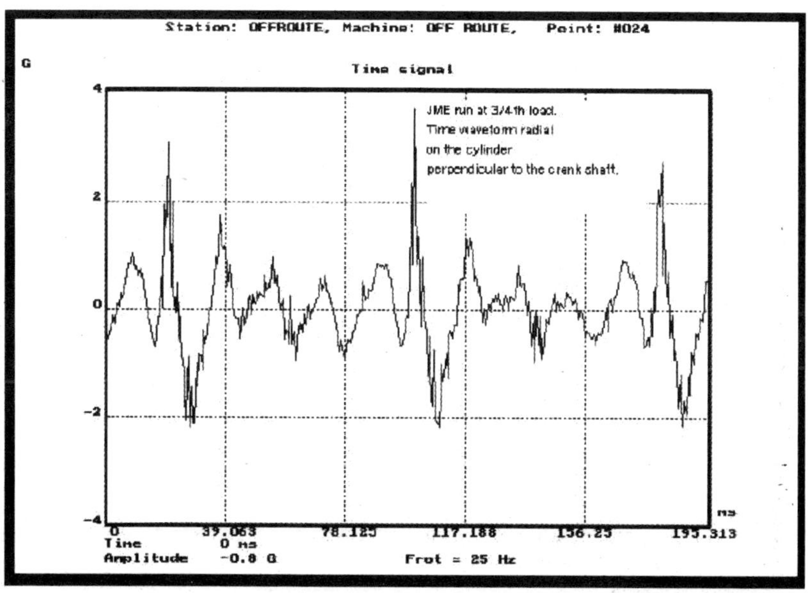

Fig:(B-16) JME run at 3/4th load. Time waveform radial on the cylinder perpendicular to the crankshaft.

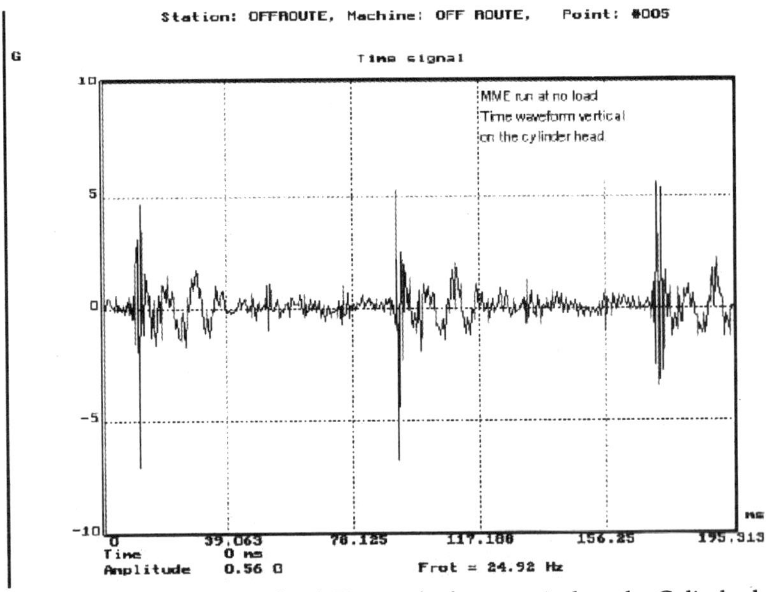

Fig:(B-17) MME run at no load. Time waveform vertical on the Cylinder head.

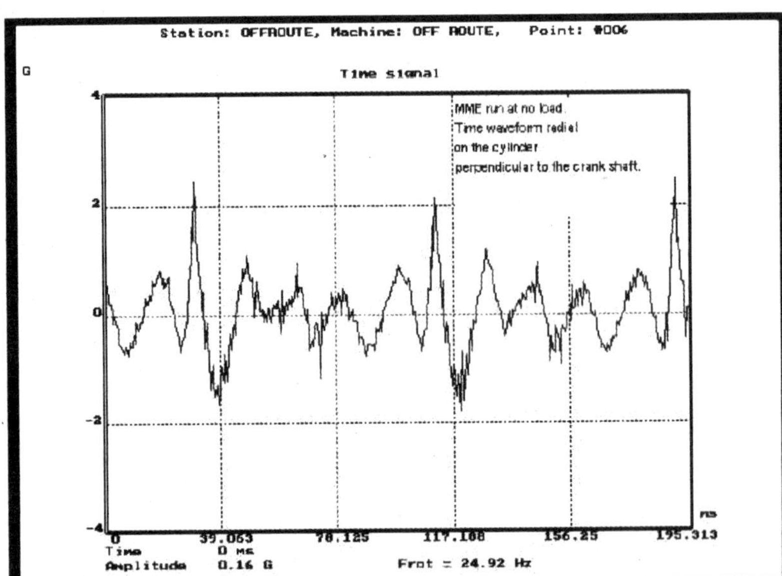

Fig:(B-18) MME run at no load. Time waveform radial on the cylinder perpendicular to the crankshaft.

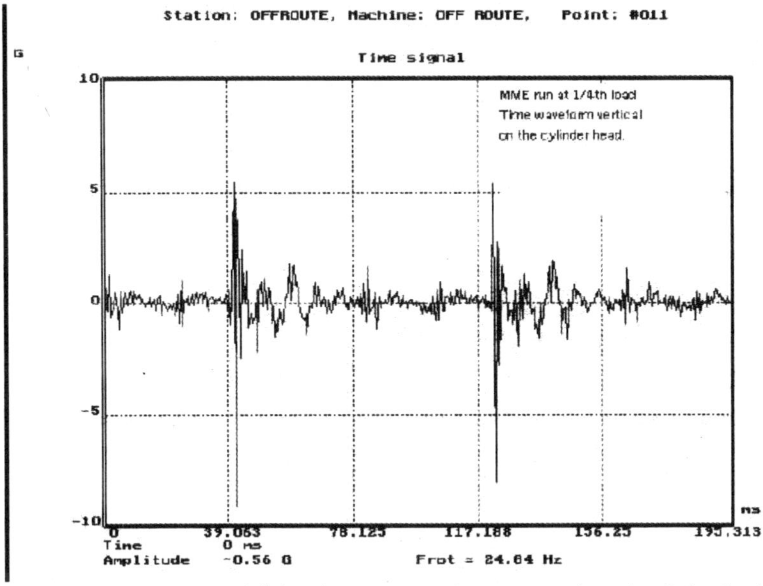

Fig:(B-19) MME run at 1/4th load. Time waveform vertical on the Cylinder head.

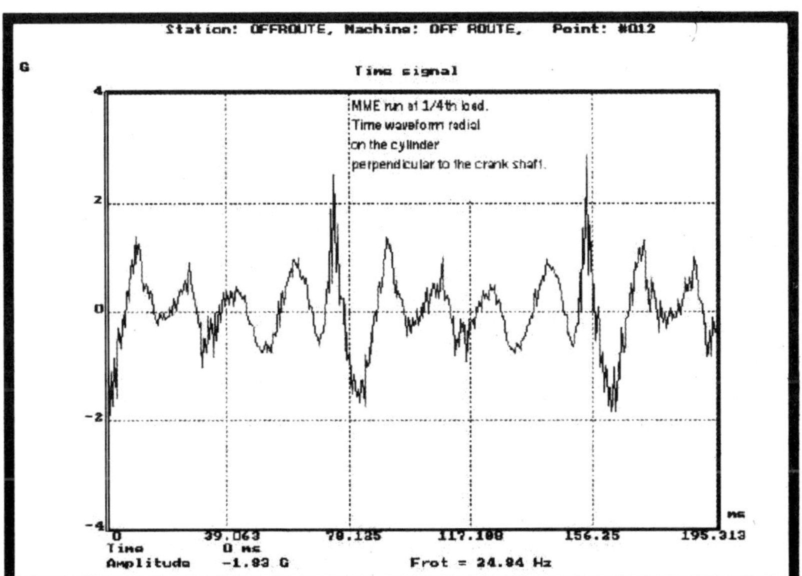

Fig:(B-20) MME run at 1/4th load. Time waveform radial on the cylinder perpendicular to the crankshaft.

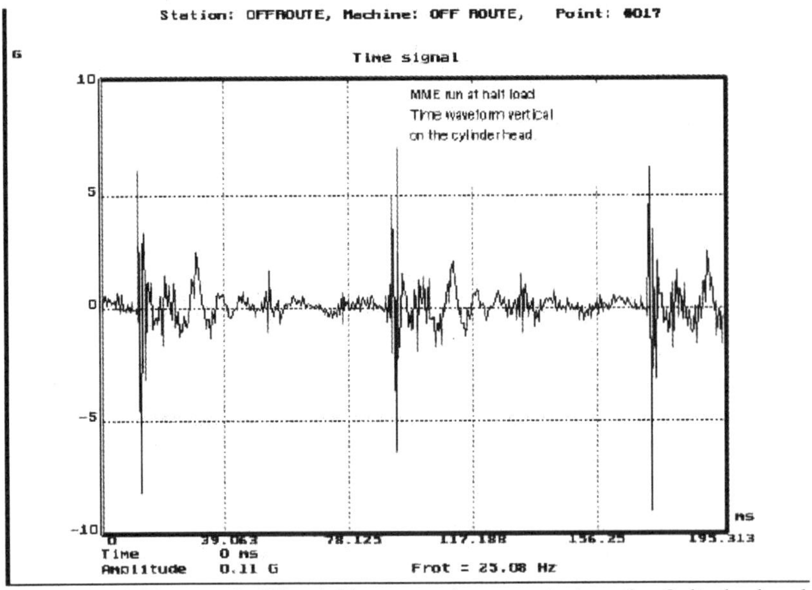

Fig:(B-21) MME run at half load. Time waveform vertical on the Cylinder head.

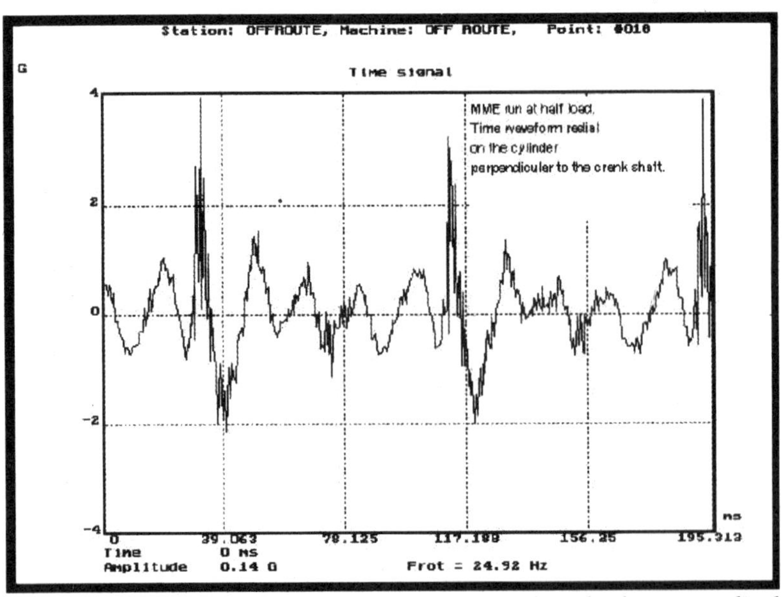

Fig:(B-22) MME run at half load. waveform radial on the cylinder perpendicular to the crankshaft.

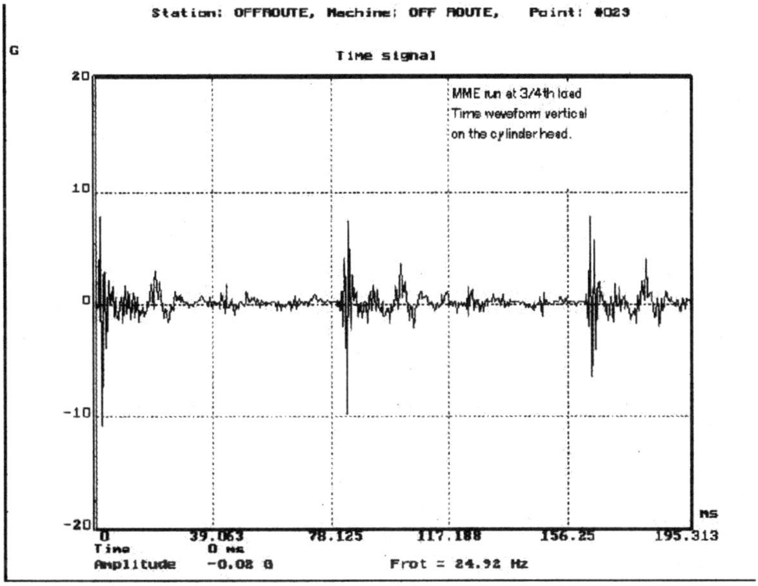

Fig:(B-23) MME run at 3/4th load. Time waveform vertical on the Cylinder head.

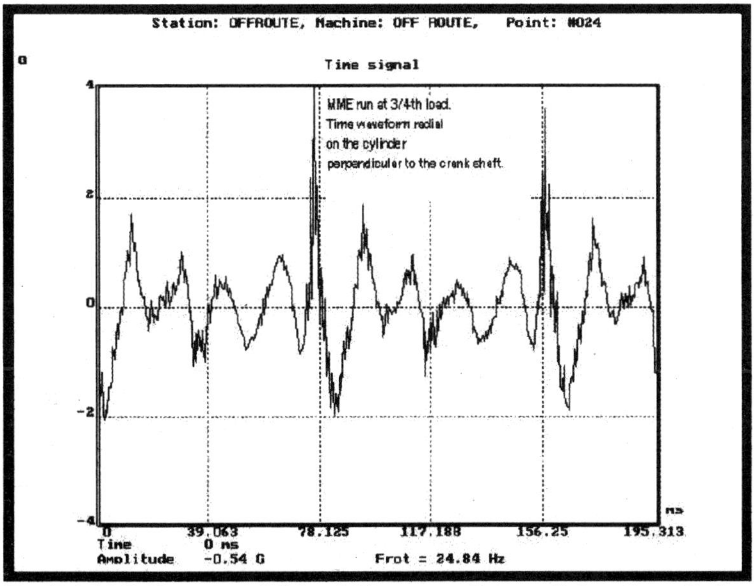

Fig:(B-24) MME run at 3/4th load. Time waveform radial on the cylinder perpendicular to the crankshaft.

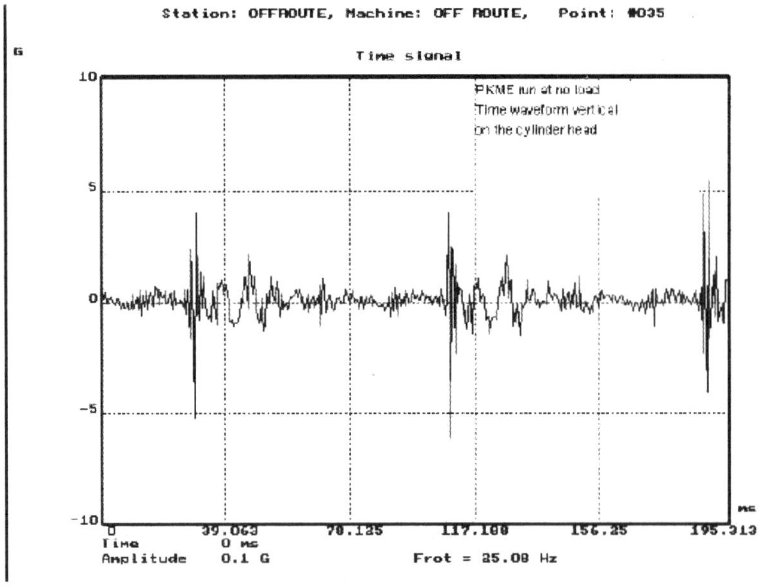

Fig:(B-25) PKME run at no load. Time waveform vertical on the Cylinder head.

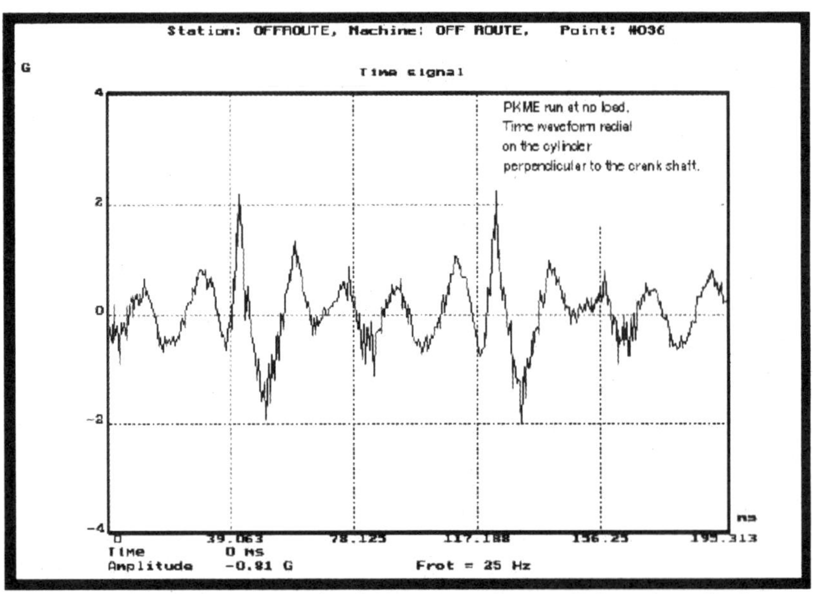

Fig:(B-26) PKME run at no load. Time waveform radial on the cylinder perpendicular to the crankshaft.

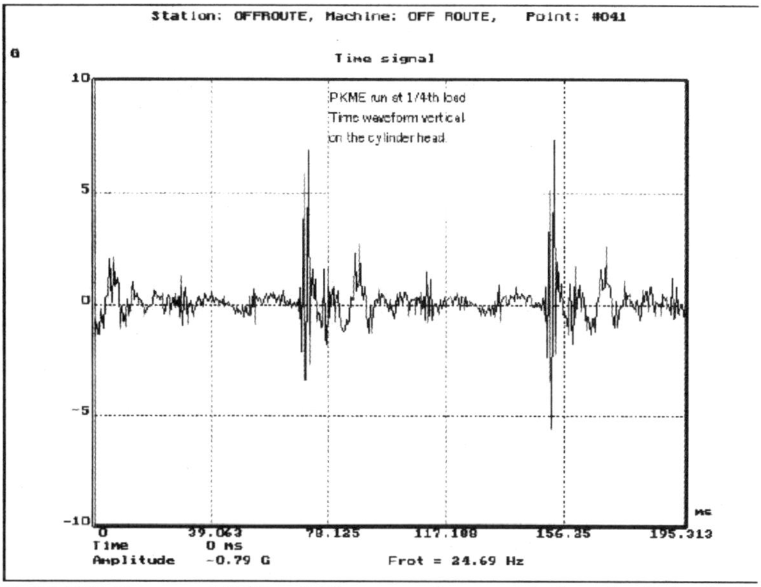

Fig:(B-27) PKME run at 1/4th load. Time waveform vertical on the Cylinder head.

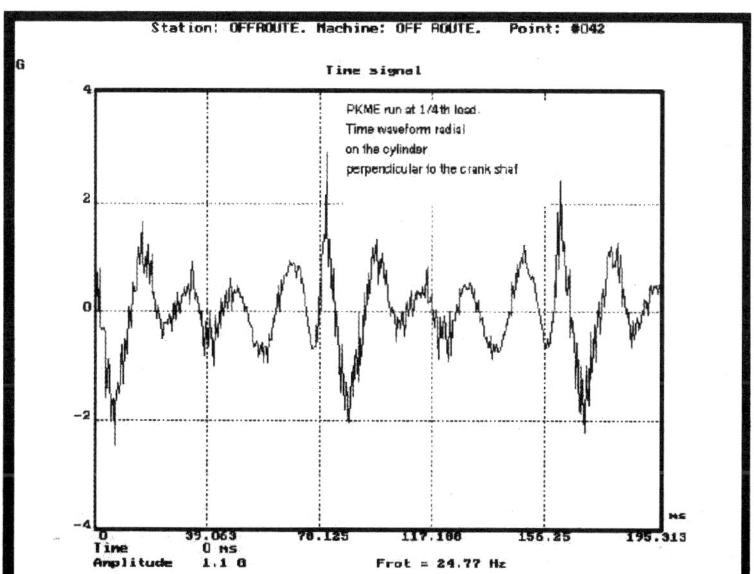

Fig:(B-28) PKME run at 1/4th load. Time waveform radial on the cylinder perpendicular to the crankshaft.

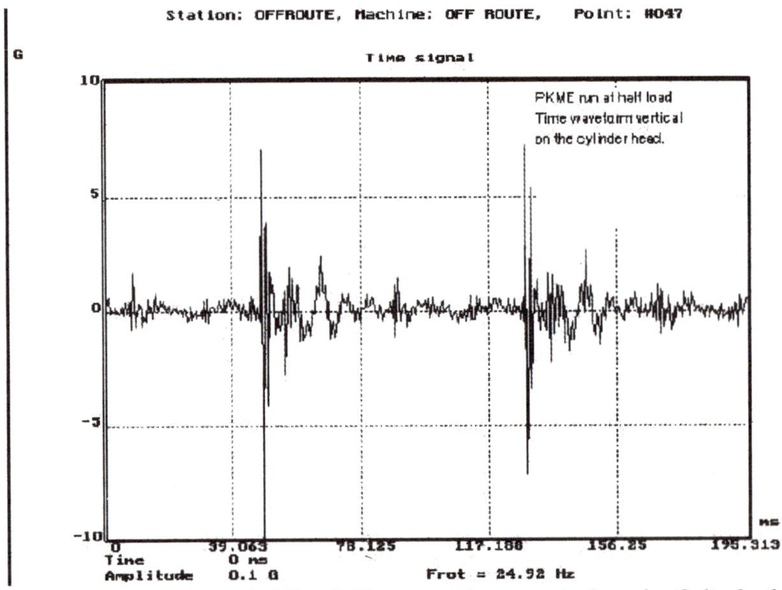

Fig:(B-29) PKME run at half load. Time waveform vertical on the Cylinder head.

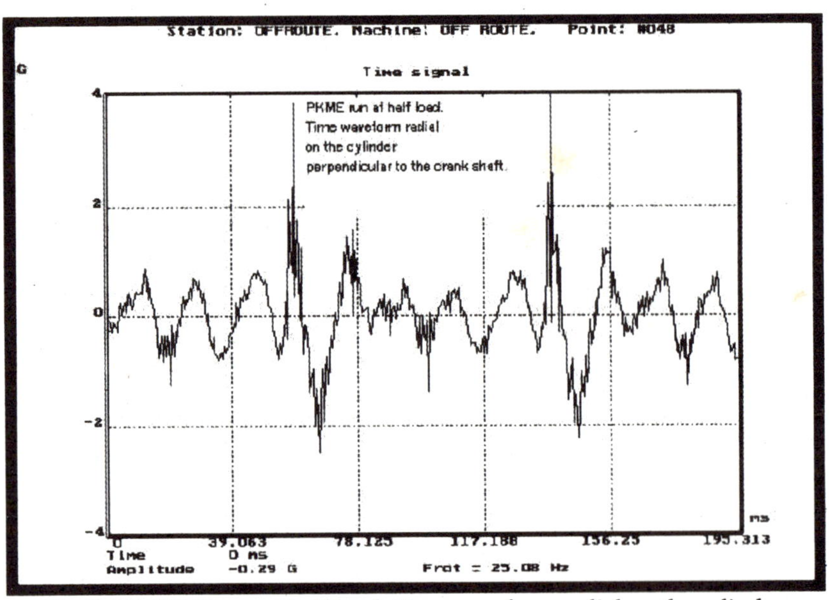

Fig:(B-30) PKME run at half load. Time waveform radial on the cylinder perpendicular to the crankshaft.

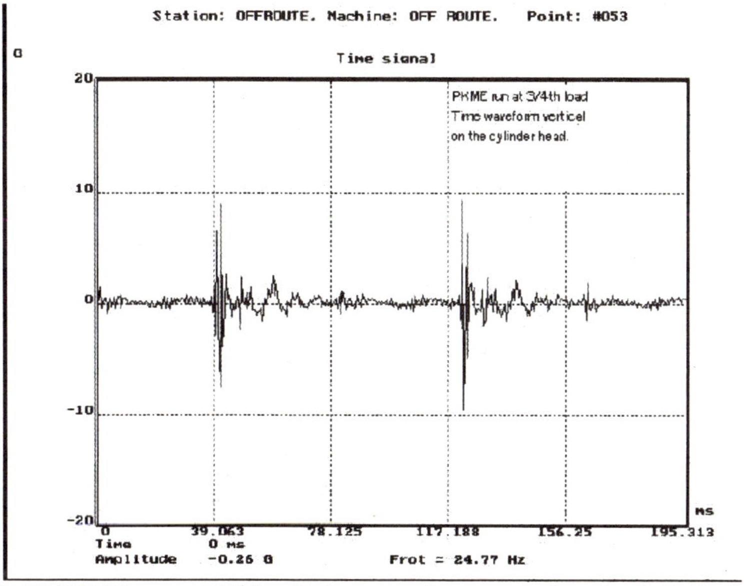

Fig:(B-31) PKME run at 3/4th load. Time waveform vertical on the Cylinder head.

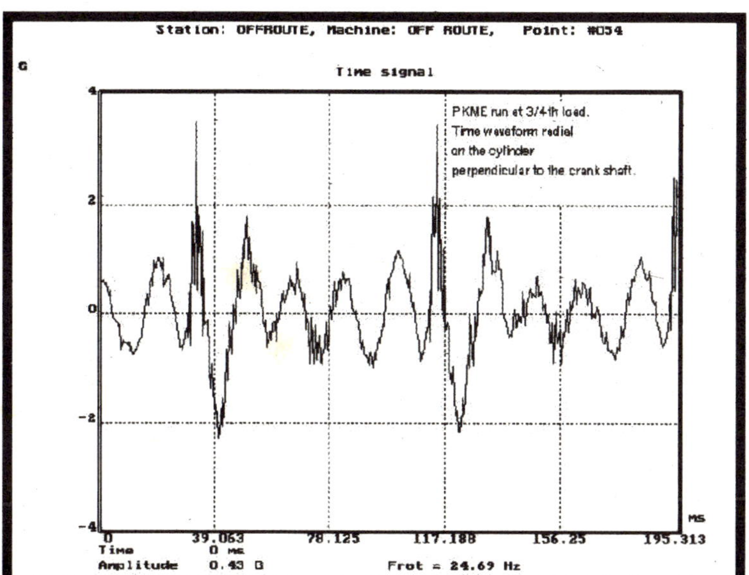

Fig:(B-32) PKME run at 3/4th load. Time waveform radial on the cylinder perpendicular to the crankshaft.

Appendix C

The pressure- crank angle diagrams for other loads not mentioned in the chapters are appended below for verification.

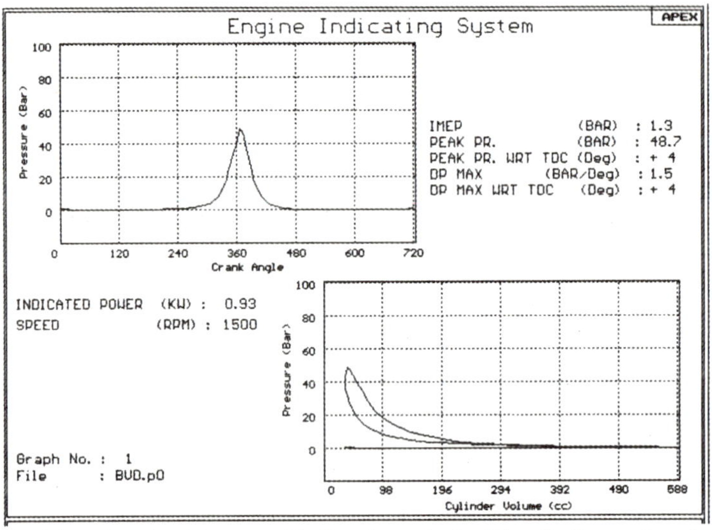

Fig:(c-1) Combustion Pressure V_s- Crank angle for Diesel oil run at no load.

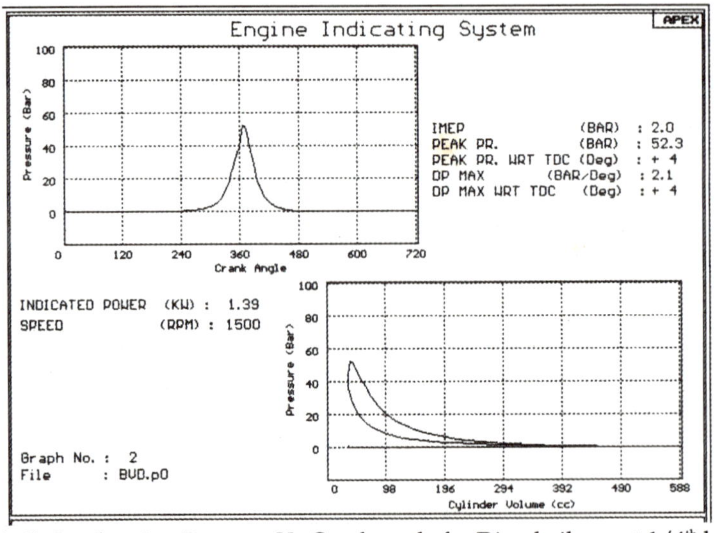

Fig:(c-2) Combustion Pressure V_s- Crank angle for Diesel oil run at 1/4th load.

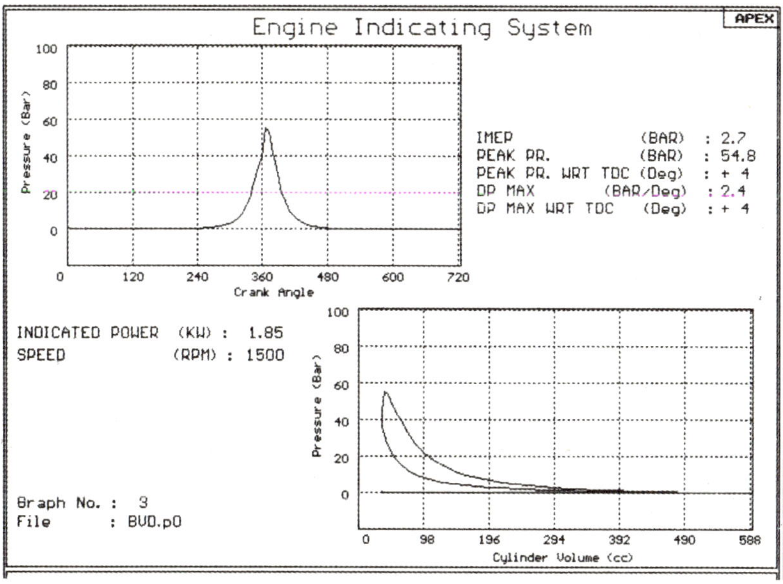

Fig:(c-3) Combustion Pressure V_{s}- Crank angle for Diesel oil run at half load.

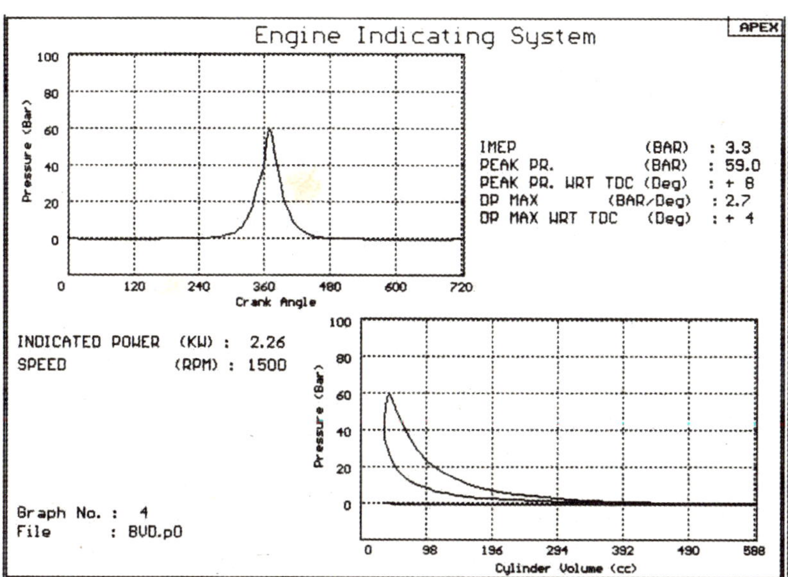

Fig:(c-4) Combustion Pressure V_{s}- Crank angle for Diesel oil run at $3/4^{th}$ load.

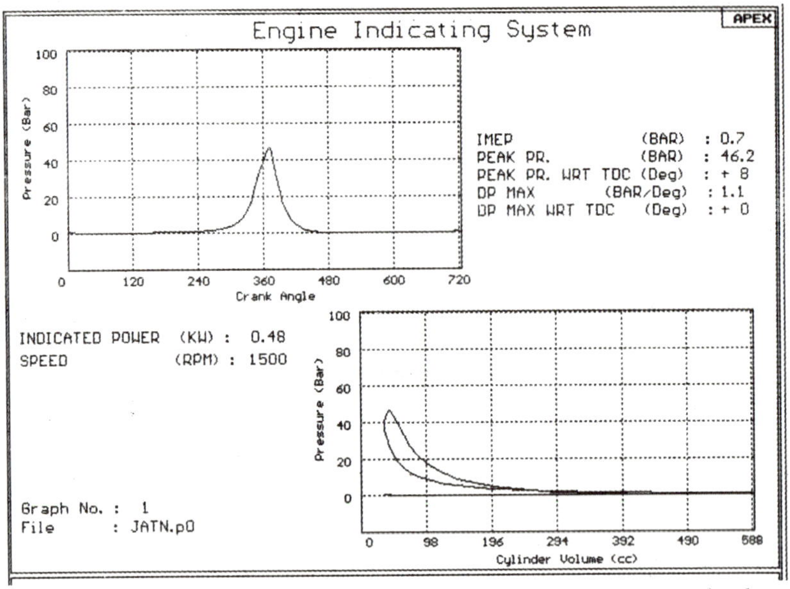

Fig:(c-5) Combustion Pressure V_s- Crank angle for JME run at no load.

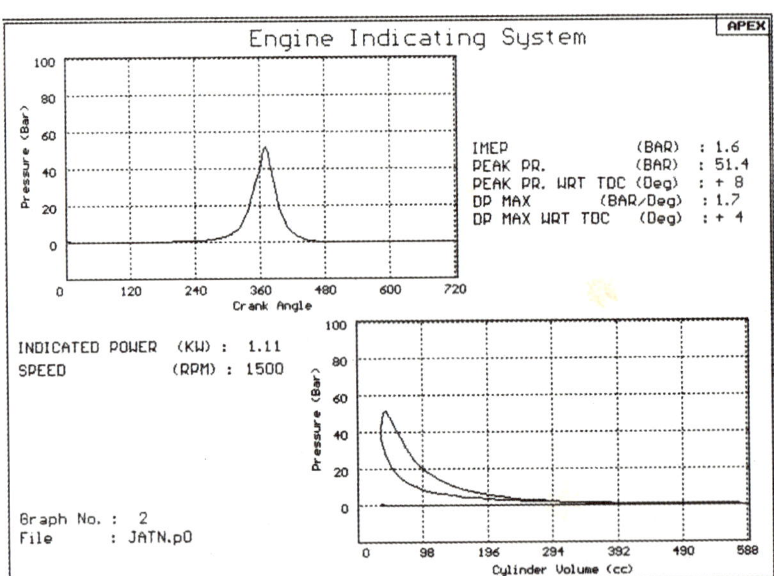

Fig:(c-6) Combustion Pressure V_s- Crank angle for JME run at 1/4th load.

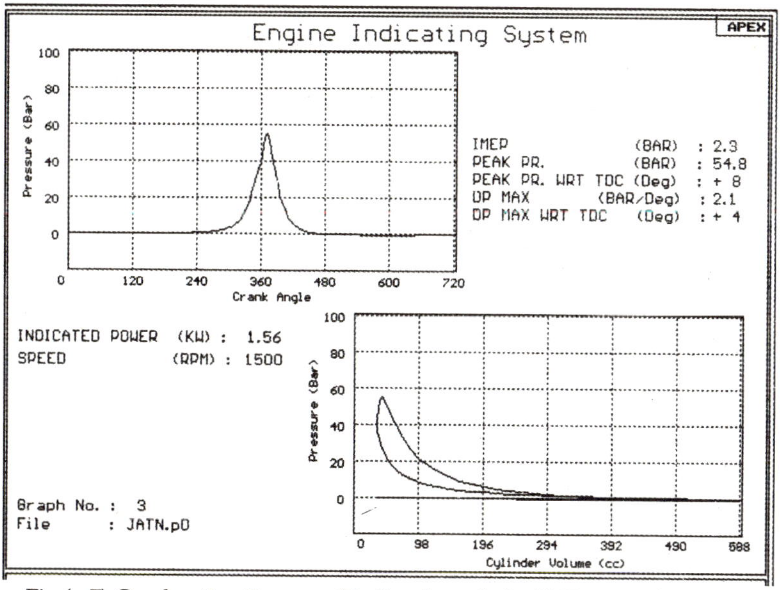

Fig:(c-7) Combustion Pressure V_{s}- Crank angle for JME run at half load.

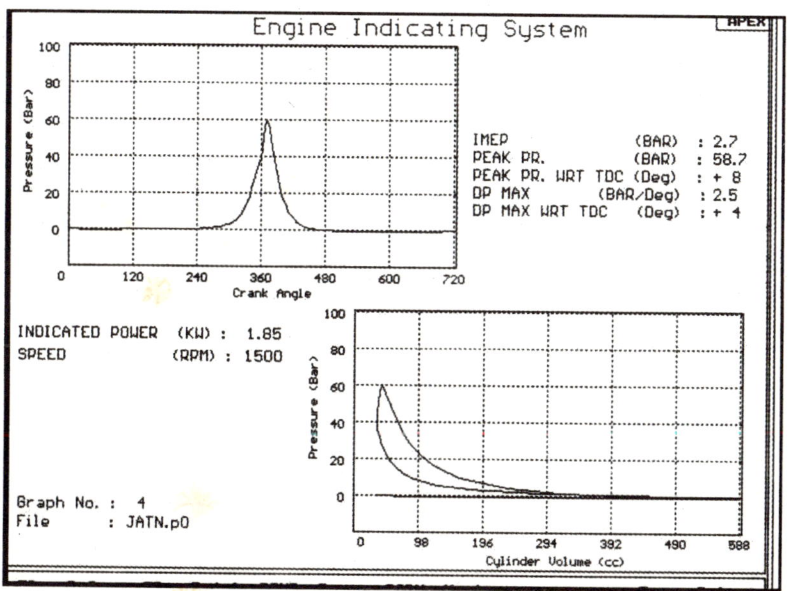

Fig:(c-8) Combustion Pressure V_{s}- Crank angle for JME run at $3/4^{th}$ load

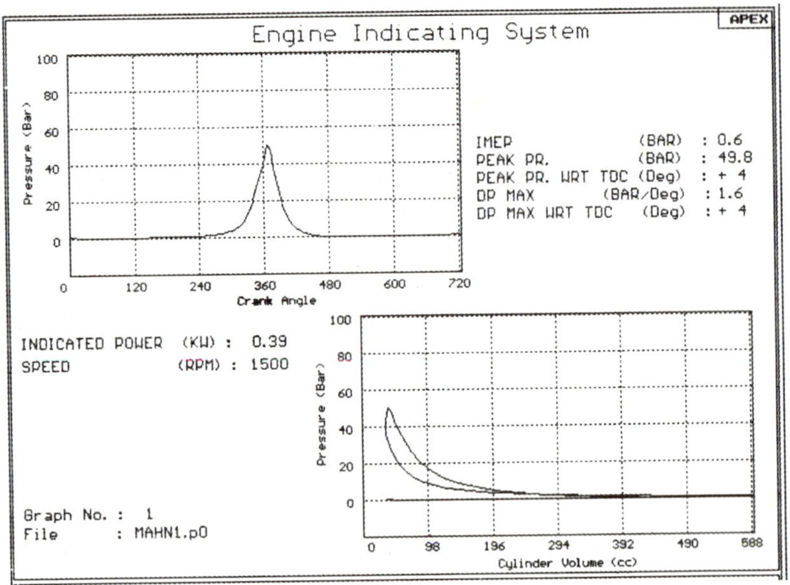

Fig:(c-9) Combustion Pressure V_s- Crank angle for MME run at no load.

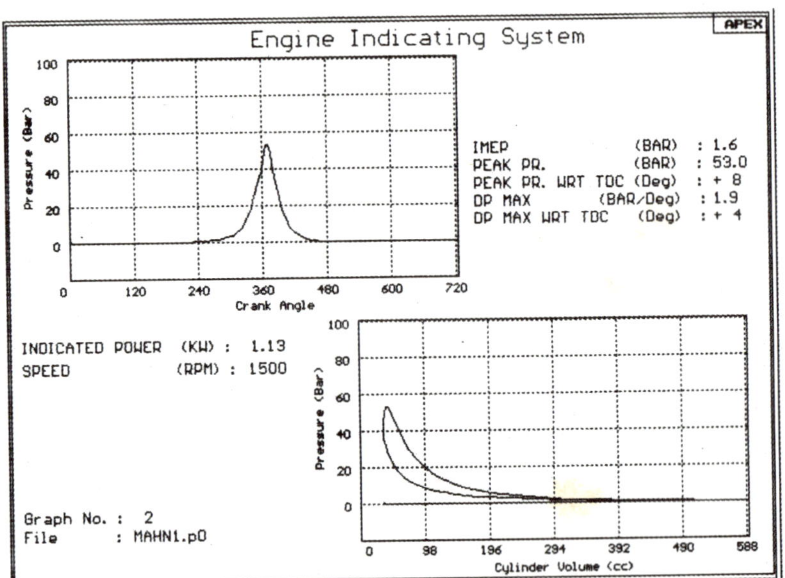

Fig:(c-10) Combustion Pressure V_s- Crank angle for MME run at $1/4^{th}$ load

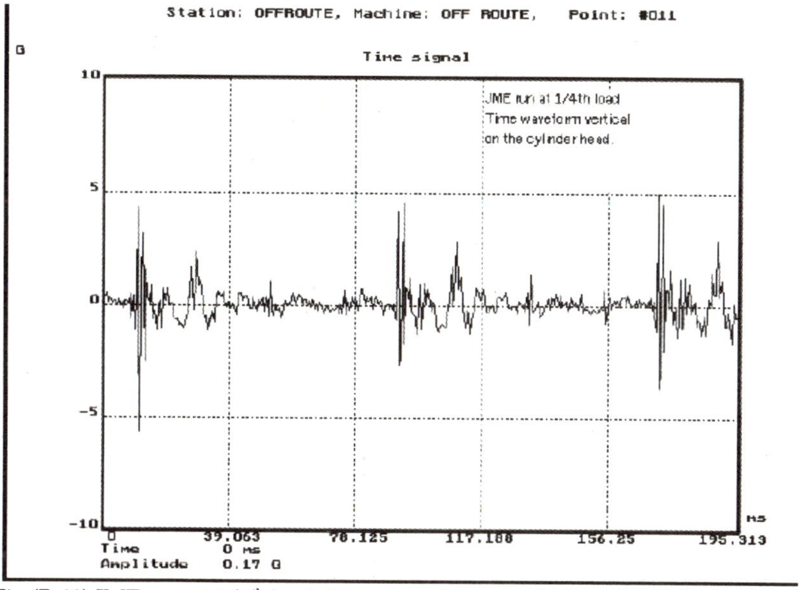

Fig:(B-11) JME run at 1/4th load. Time waveform vertical on the Cylinder head.

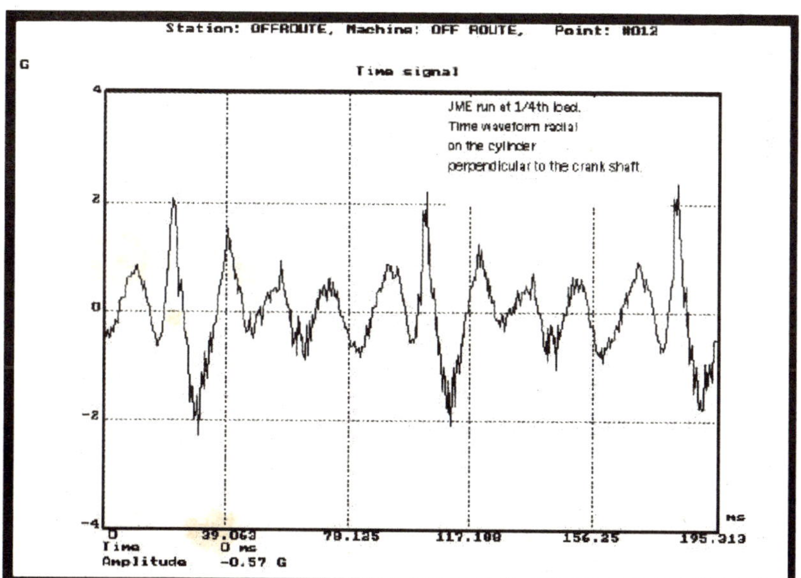

Fig:(B-12) JME run at 1/4th load. Time waveform radial on the cylinder perpendicular to the crankshaft.

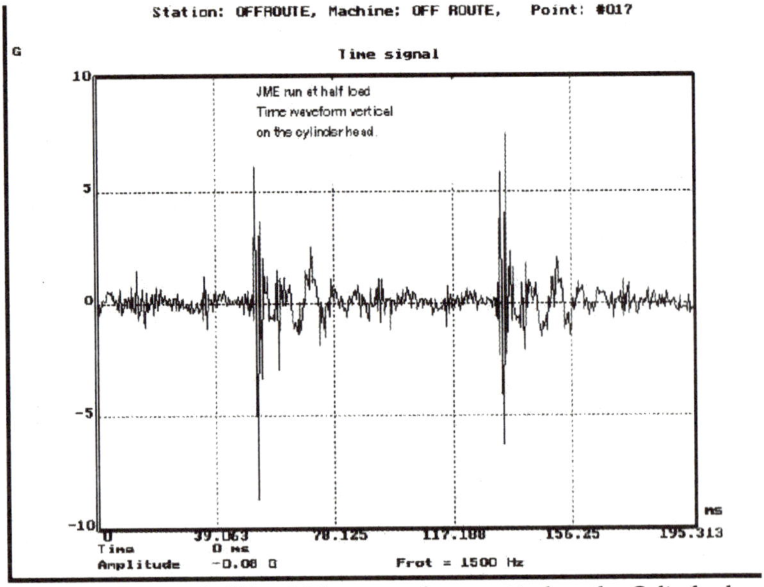

Fig:(B-13) JME run at half load. Time waveform vertical on the Cylinder head.

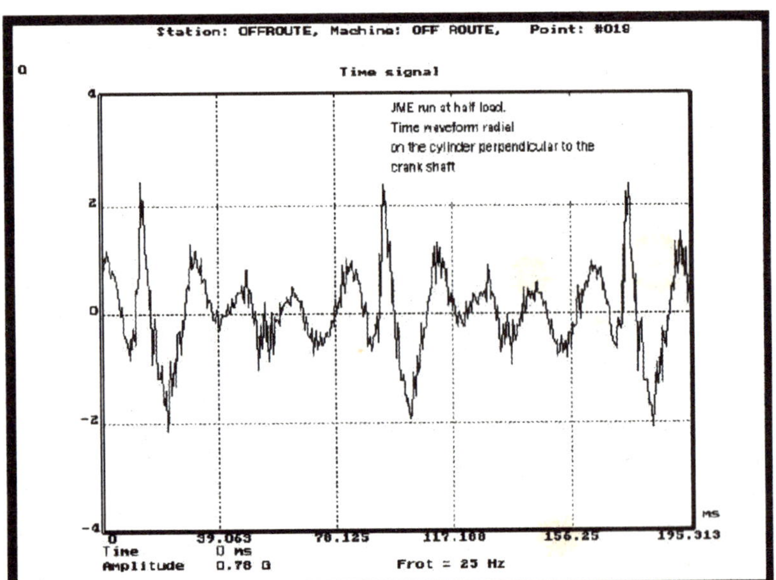

Fig:(B-14) JME run at half load. Time waveform radial on the cylinder perpendicular to the crankshaft.

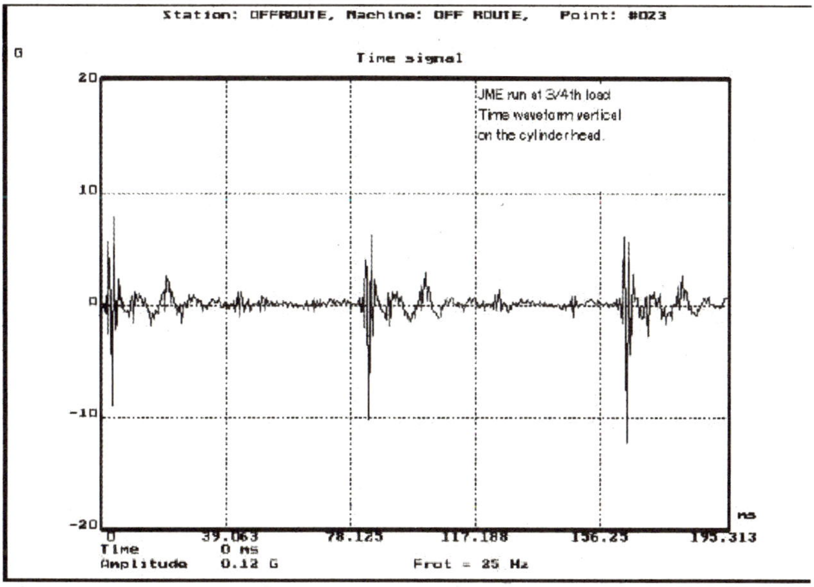

Fig:(B-15) JME run at 3/4th load. Time waveform vertical on the Cylinder head.

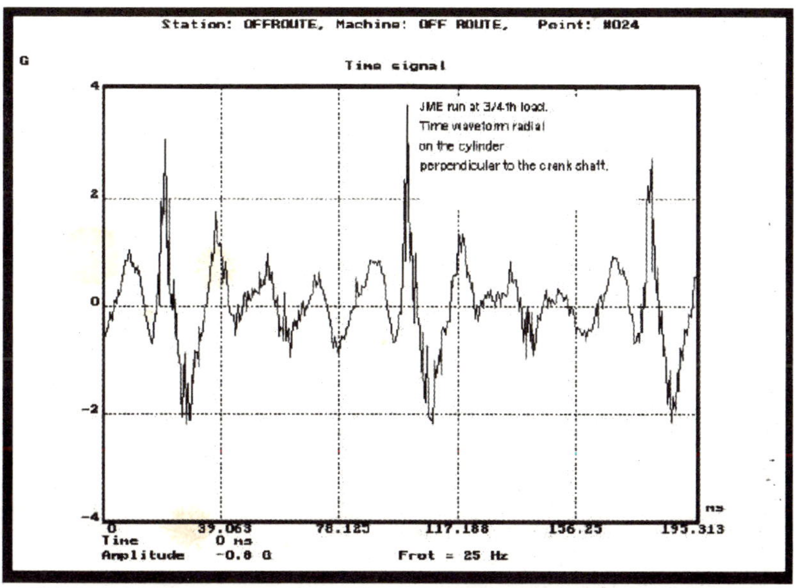

Fig:(B-16) JME run at 3/4th load. Time waveform radial on the cylinder perpendicular to the crankshaft.

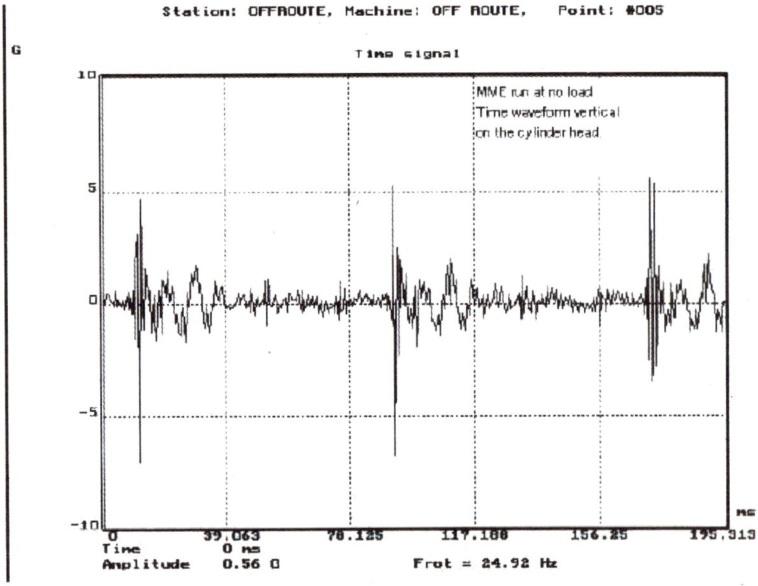

Fig:(B-17) MME run at no load. Time waveform vertical on the Cylinder head.

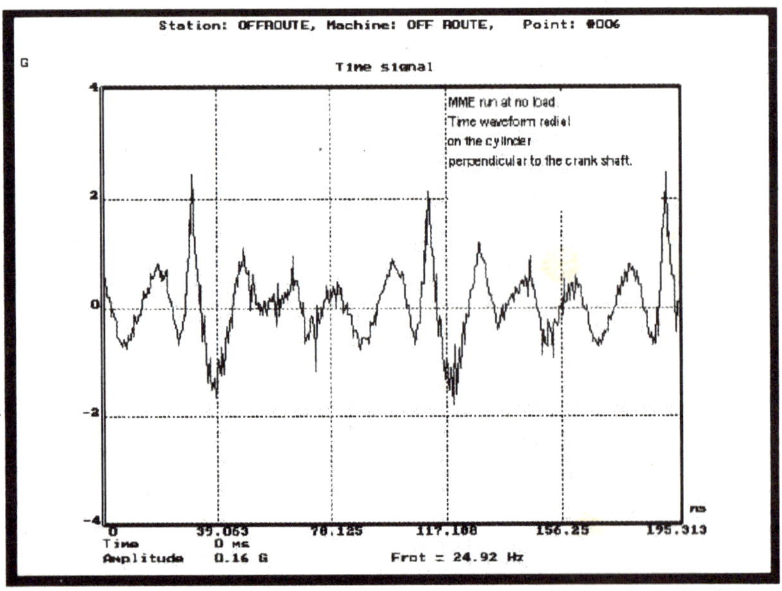

Fig:(B-18) MME run at no load. Time waveform radial on the cylinder perpendicular to the crankshaft.

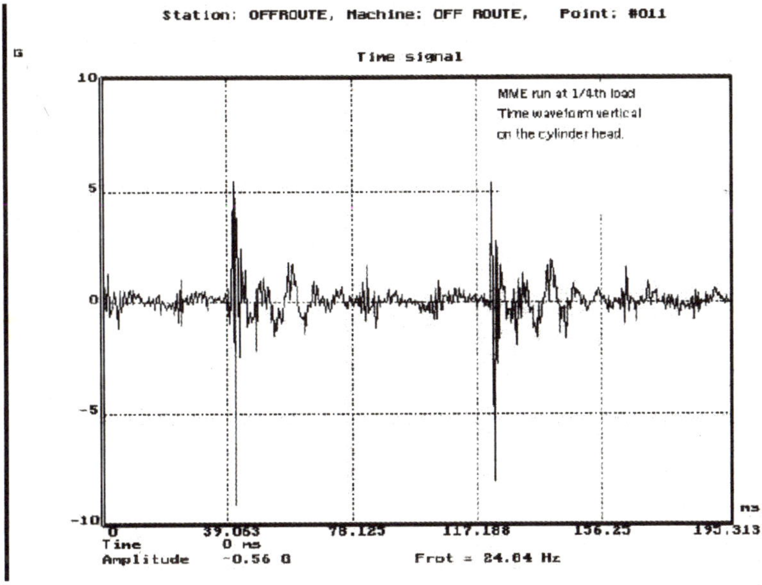

Fig:(B-19) MME run at 1/4th load. Time waveform vertical on the Cylinder head.

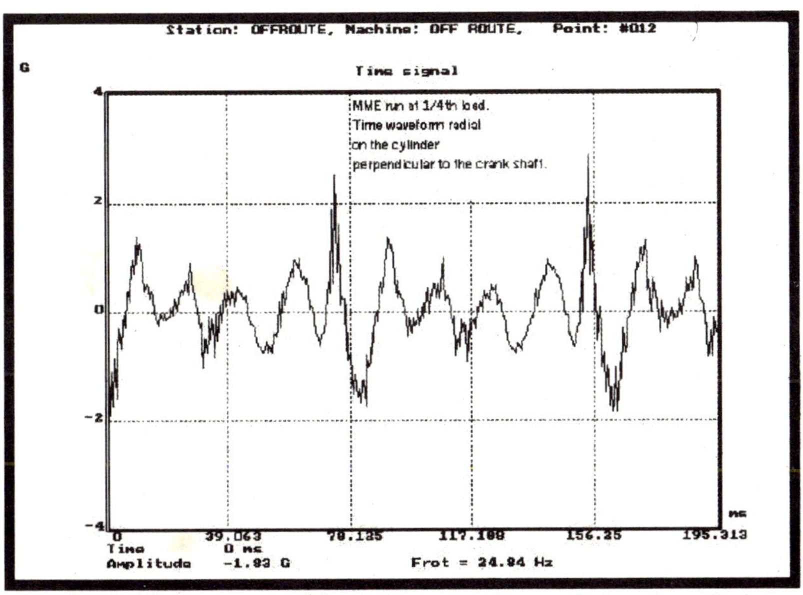

Fig:(B-20) MME run at 1/4th load. Time waveform radial on the cylinder perpendicular to the crankshaft.

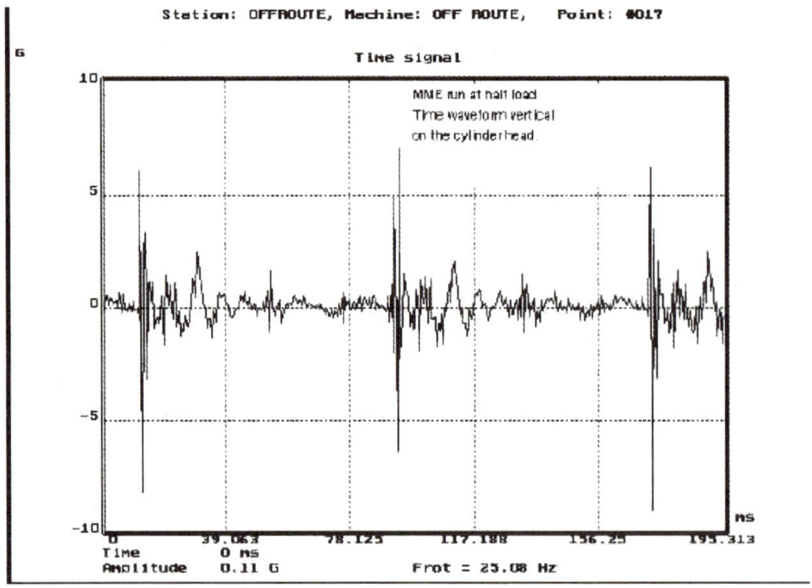

Fig:(B-21) MME run at half load. Time waveform vertical on the Cylinder head.

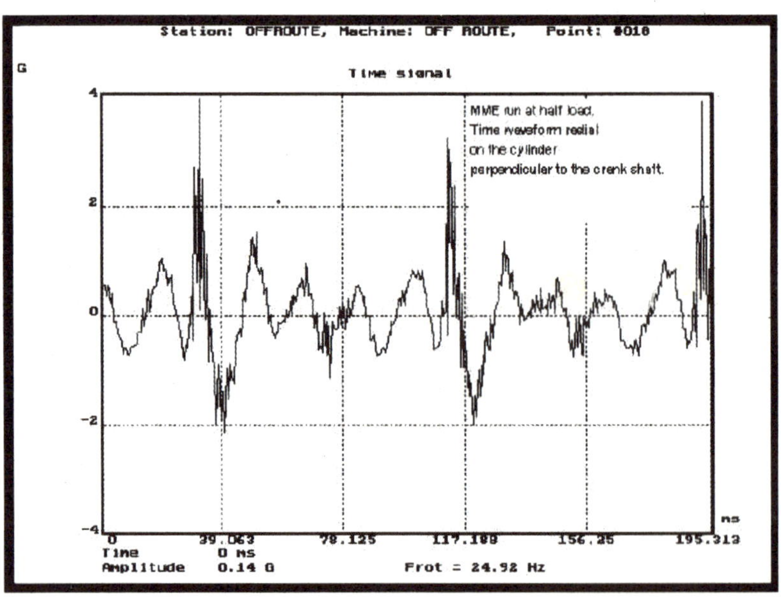

Fig:(B-22) MME run at half load. waveform radial on the cylinder perpendicular to the crankshaft.

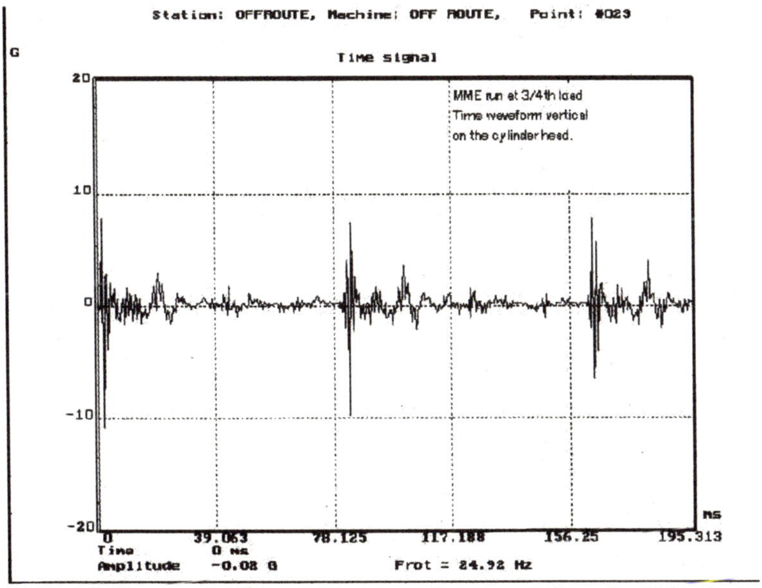

Fig:(B-23) MME run at 3/4th load. Time waveform vertical on the Cylinder head.

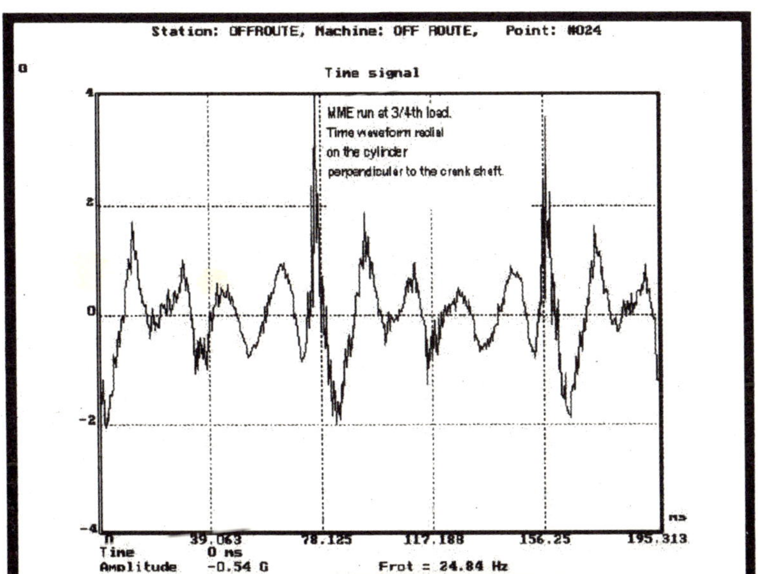

Fig:(B-24) MME run at 3/4th load. Time waveform radial on the cylinder perpendicular to the crankshaft.

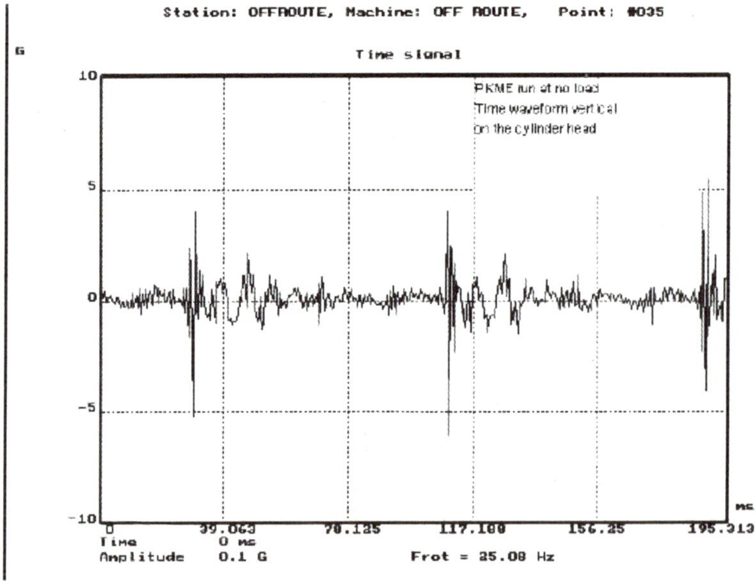

Fig:(B-25) PKME run at no load. Time waveform vertical on the Cylinder head.

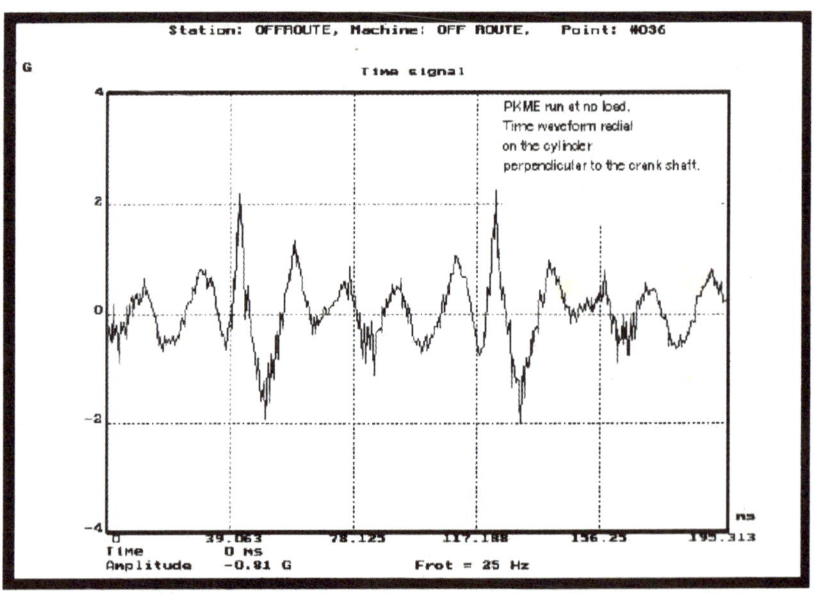

Fig:(B-26) PKME run at no load. Time waveform radial on the cylinder perpendicular to the crankshaft.

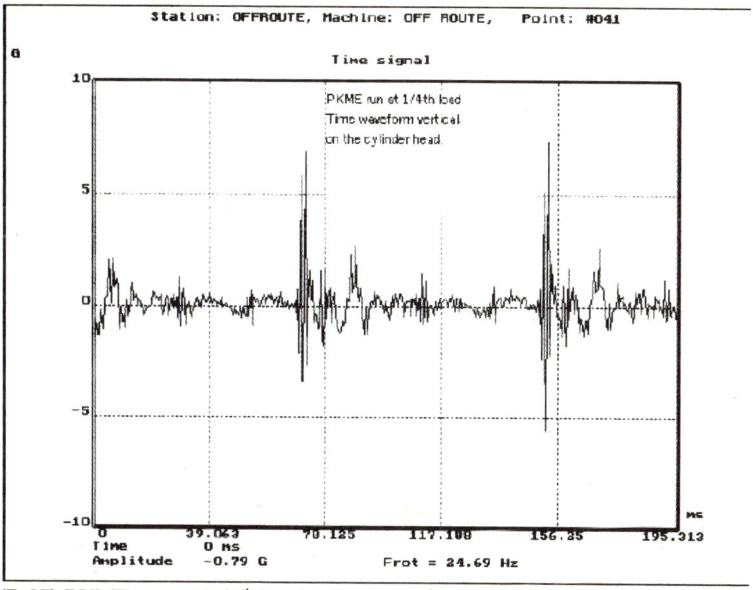

Fig:(B-27) PKME run at 1/4th load. Time waveform vertical on the Cylinder head.

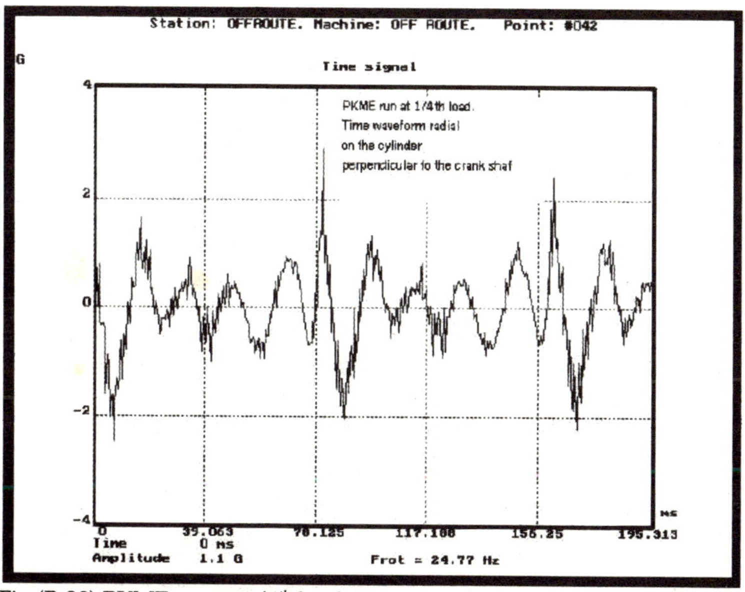

Fig:(B-28) PKME run at 1/4th load. Time waveform radial on the cylinder perpendicular to the crankshaft.

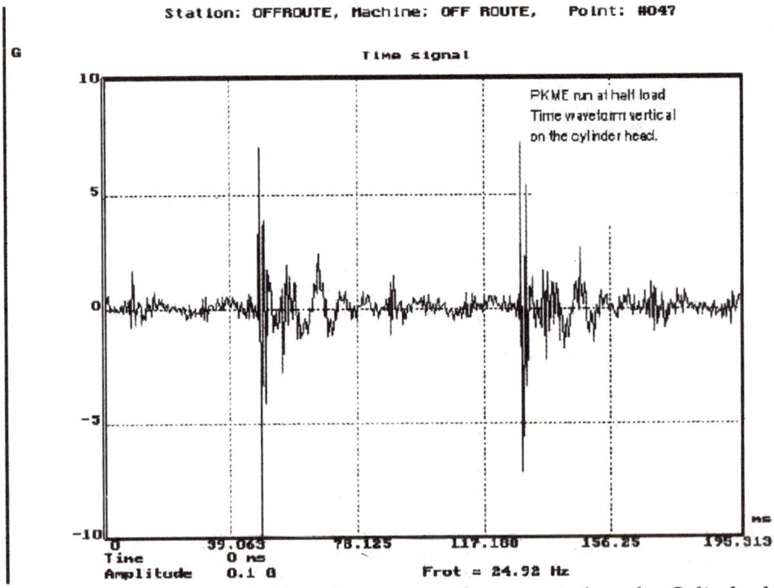

Fig:(B-29) PKME run at half load. Time waveform vertical on the Cylinder head.

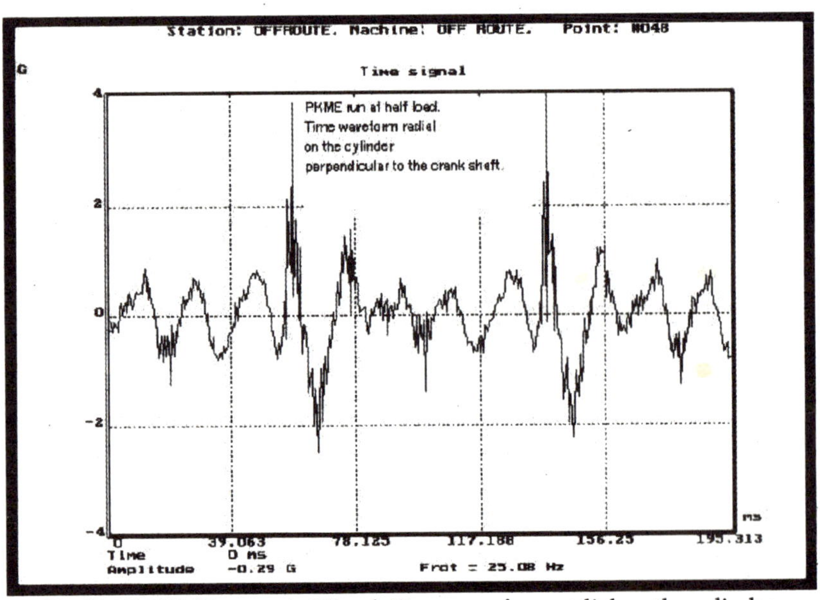

Fig:(B-30) PKME run at half load. Time waveform radial on the cylinder perpendicular to the crankshaft.

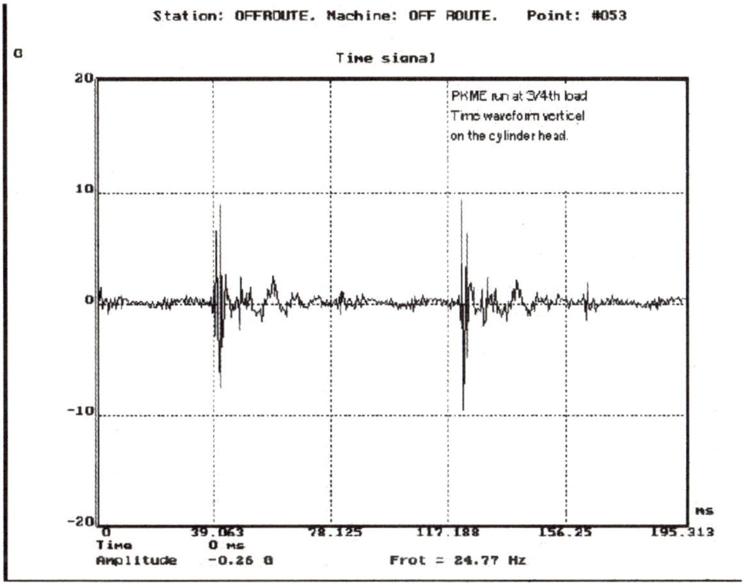

Fig:(B-31) PKME run at 3/4th load. Time waveform vertical on the Cylinder head.

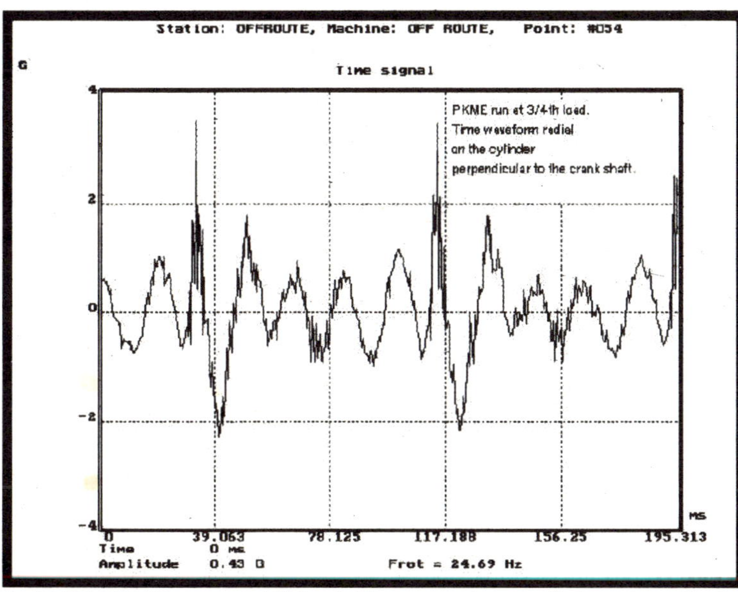

Fig:(B-32) PKME run at 3/4th load. Time waveform radial on the cylinder perpendicular to the crankshaft.

Appendix C

The pressure- crank angle diagrams for other loads not mentioned in the chapters are appended below for verification.

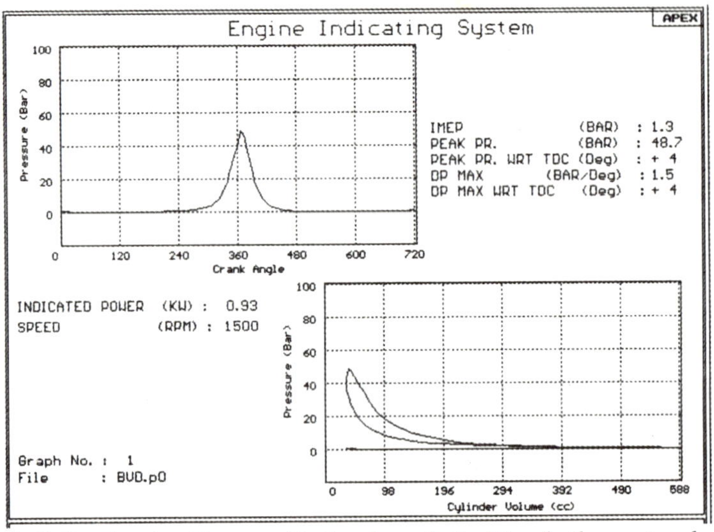

Fig:(c-1) Combustion Pressure V_{s}- Crank angle for Diesel oil run at no load.

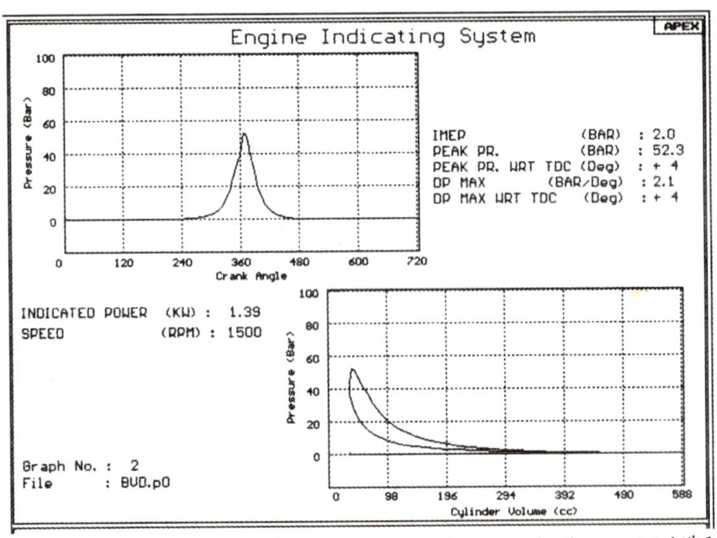

Fig:(c-2) Combustion Pressure V_{s}- Crank angle for Diesel oil run at $1/4^{th}$ load.

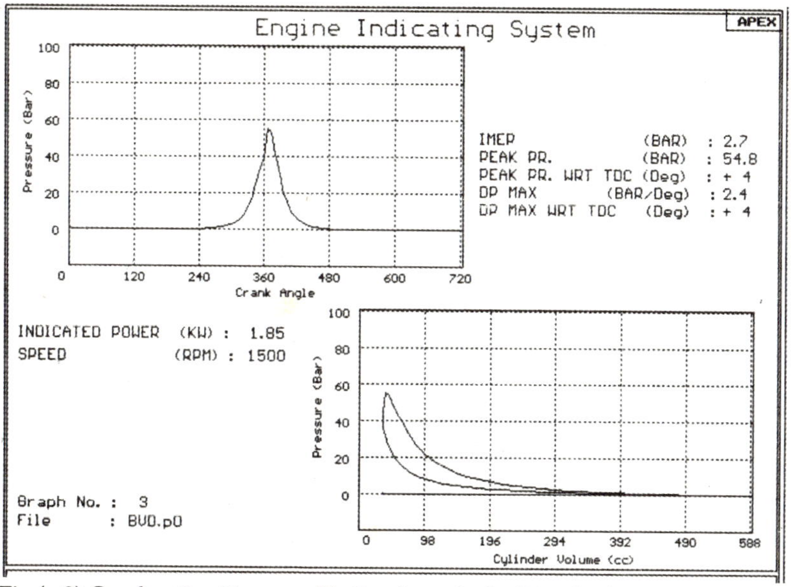

Fig:(c-3) Combustion Pressure Vs- Crank angle for Diesel oil run at half load.

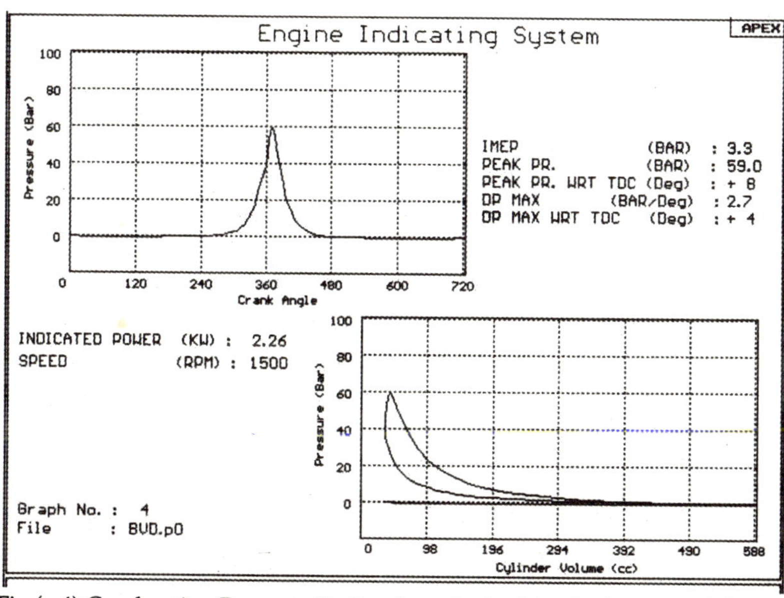

Fig:(c-4) Combustion Pressure Vs- Crank angle for Diesel oil run at 3/4th load.

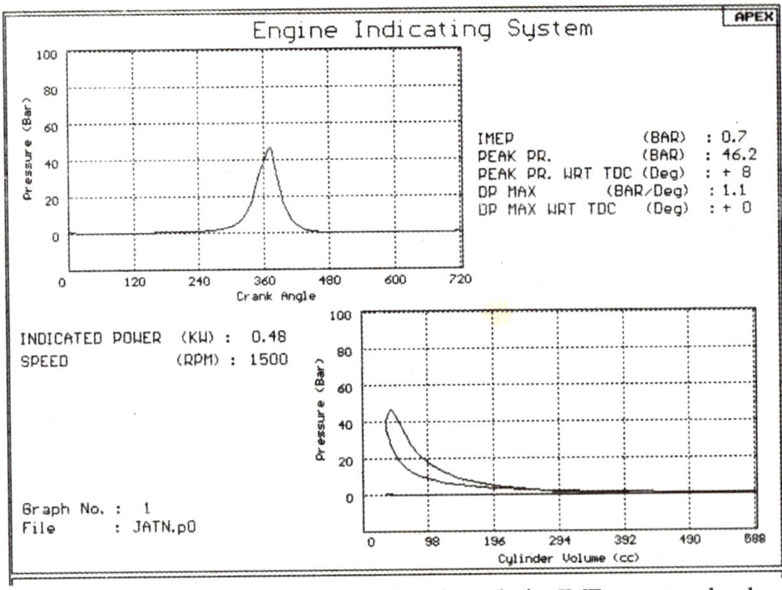

Fig:(c-5) Combustion Pressure Vs- Crank angle for JME run at no load.

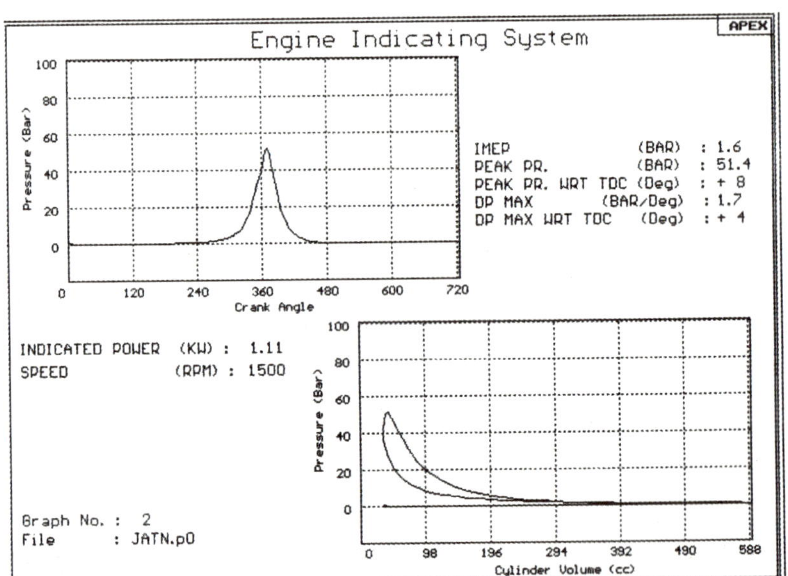

Fig:(c-6) Combustion Pressure Vs- Crank angle for JME run at 1/4th load.

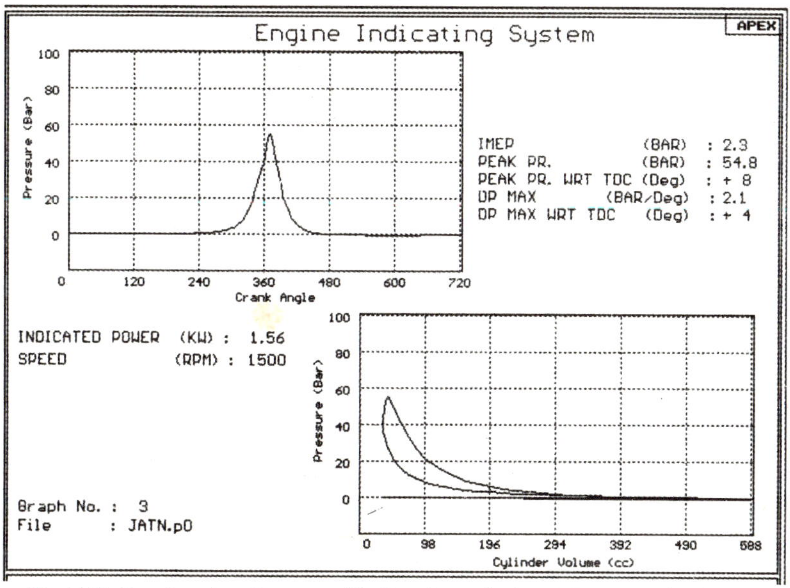

Fig:(c-7) Combustion Pressure V_{s}- Crank angle for JME run at half load.

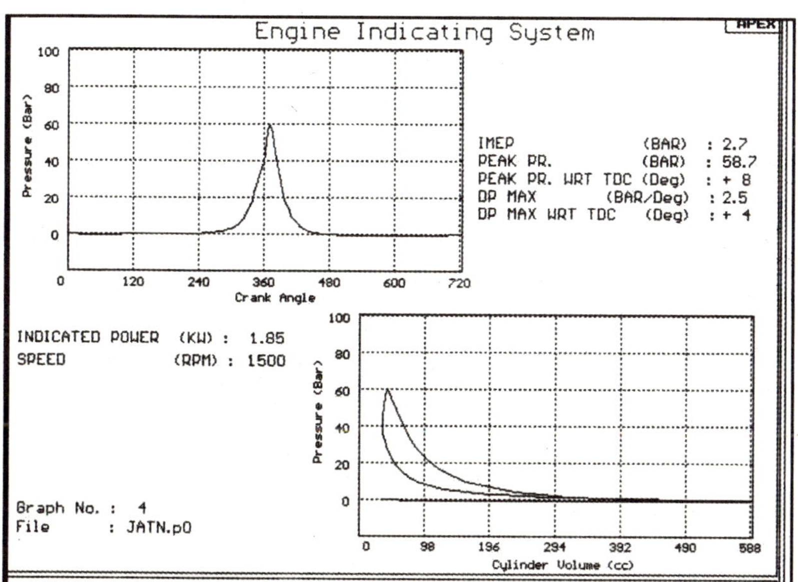

Fig:(c-8) Combustion Pressure V_{s}- Crank angle for JME run at 3/4th load

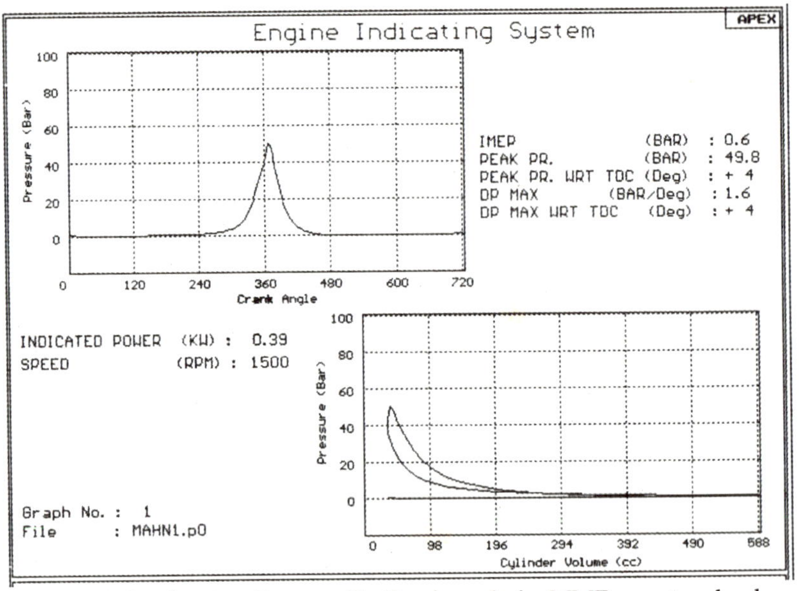

Fig:(c-9) Combustion Pressure Vs- Crank angle for MME run at no load.

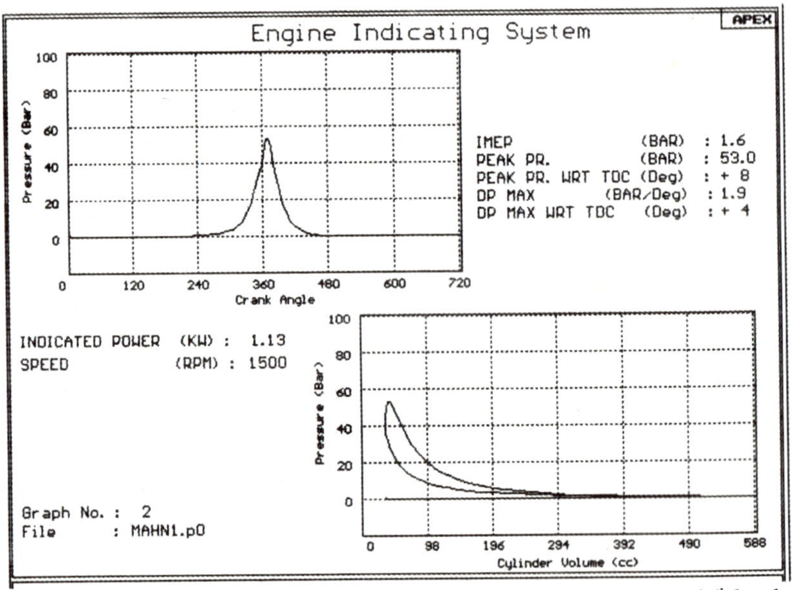

Fig:(c-10) Combustion Pressure Vs- Crank angle for MME run at 1/4th load

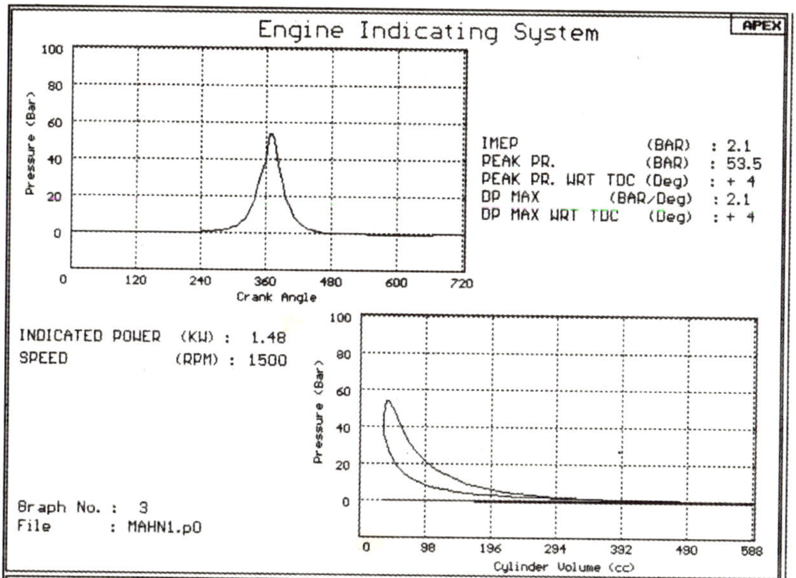

Fig:(c-11) Combustion Pressure $V_{s\text{-}}$ Crank angle for MME run at half load.

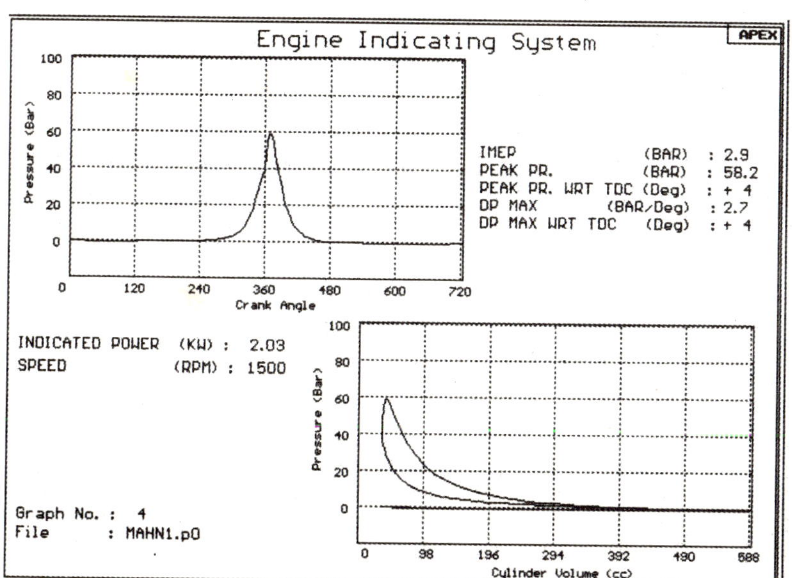

Fig:(c-12) Combustion Pressure $V_{s\text{-}}$ Crank angle for MME run at $3/4^{th}$ load.

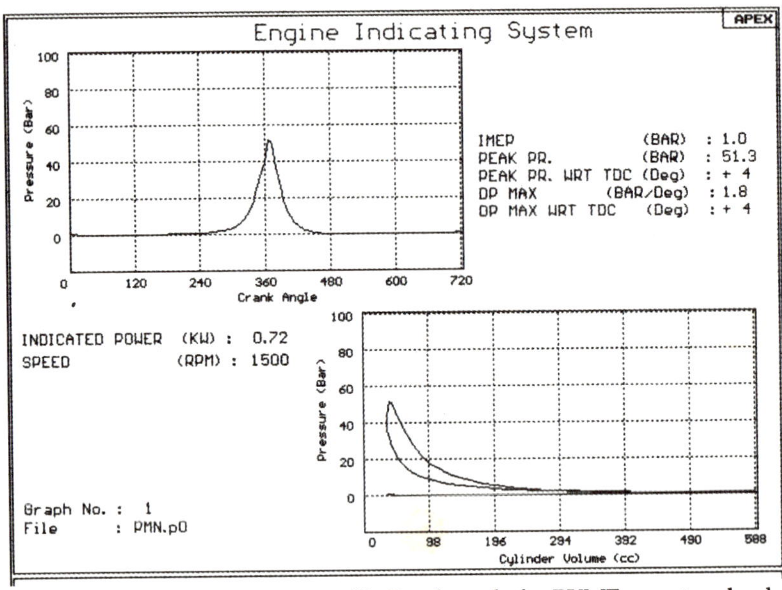

Fig:(c-13) Combustion Pressure V_{s}- Crank angle for PKME run at no load.

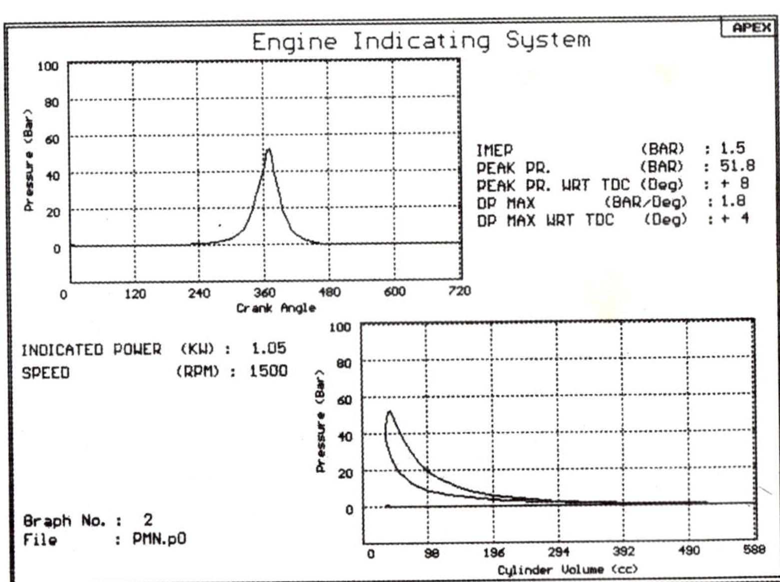

Fig:(c-14) Combustion Pressure V_{s}- Crank angle for PKME run at $1/4^{th}$ load.

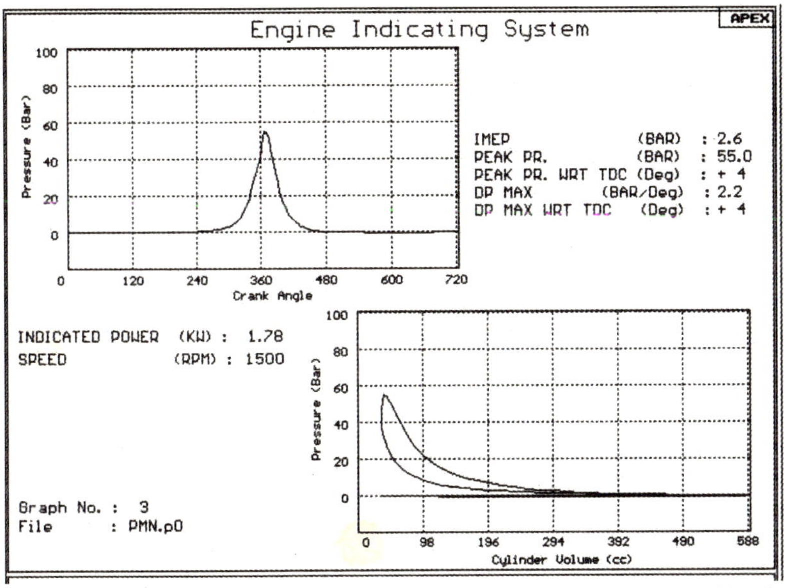

Fig:(c-15) Combustion Pressure V_{s}- Crank angle for PKME run at half load.

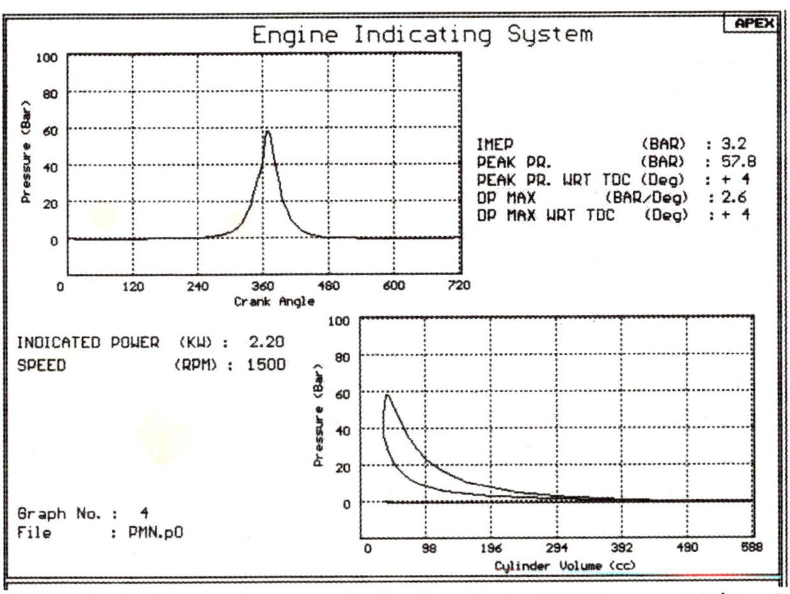

Fig: (c-16) Combustion Pressure V_{s}- Crank angle for PKME run at 3/4th load.

Index